CHROMATOGRAPHY FOR INORGANIC CHEMISTRY

CHROMATOGRAPHY FOR INORGANIC CHEMISTRY

Michael Lederer
Institut de Chimie Minèrale et Analytique,
Université de Lausanne, Switzerland

JOHN WILEY & SONS
Chichester · New York · Brisbane · Toronto · Singapore

Copyright © 1994 by John Wiley & Sons Ltd,
Baffins Lane, Chichester,
West Sussex PO19 1UD, England
Telephone (+44) (0) 243 779777

All rights reserved.

No part of this book may be reproduced by any means, or transmitted, or translated into a machine language without the written permission of the publisher.

Other Wiley Editorial Offices

John Wiley & Sons, Inc., 605 Third Avenue,
New York, NY 10158-0012, USA

Jacaranda Wiley Ltd, 33 Park Road, Milton,
Queensland 4065, Australia

John Wiley & Sons (Canada) Ltd, 22 Worcester Road,
Rexdale, Ontario M9W 1L1, Canada

John Wiley & Sons (SEA) Pte Ltd, 37 Jalan Pemimpin #05-04,
Block B, Union Industrial Building, Singapore 2057

Library of Congress Cataloging-in-Publication Data

Lederer, Michael, 1924–
    Chromatography for inorganic chemistry / Michael Lederer.
      p.   cm.
    Includes bibliographical references and index.
    ISBN 0 471 94285 5: — ISBN 0 471 94286 3 (pbk.)
    1. Chromatographic analysis. 2. Inorganic compounds. I. Title.
QD79. C4L44   1994
543'.089 — dc20   93–31657
                                                      CIP

British Library Cataloguing in Publication Data

A catalogue record for this book is available from the British Library

ISBN 0 471 94285 5 (cloth)
ISBN 0 471 94286 3 (paper)

Printed and bound in Great Britain by Biddles Ltd, Guildford, Surrey

CONTENTS

Preface — vii

1 Historical Introduction — 1

2 Solvent Extraction and the History of Partition Chromatography — 10

3 Paper Chromatography and Thin Layer Chromatography — 28

4 Electrophoresis — 50

5 Gel Filtration — 80

6 Ion Exchange — 86

7 HPLC — 118

8 Ion Chromatography — 126

9 Gas Chromatography — 133

10 Separation of Isotopes — 140

11 Separation of Optical Isomers — 151

12 Some Elements and their Chromatography and Electrophoresis — 162

   (i) Boron — 162

   (ii) Condensed phosphates — 163

   (iii) Sulphur compounds — 170

   (iv) Halogen acids, especially halogen oxyacids — 176

   (v) Rhenium and technetium — 180

   (vi) Ruthenium — 187

(vii)	Rhodium	193
(viii)	The rare earths	199
(ix)	Chromatography at tracer levels	208
	Polonium	209
	Protactinium	211
Index		219

PREFACE

This book is the textbook for a course of lectures on chromatography in inorganic chemistry. It assumes that the student is familiar with the principles of chromatography from his course in analytical chemistry, and it sets out to show the advances made in inorganic chemistry, especially in the chemistry of aqueous solutions, by using chromatographic and electrophoretic methods.

I have given this course to advanced students at the University of Lausanne in 1981, at the Hebrew University of Jerusalem in 1989 and at the Ben Gurion University of the Negev in 1991. As it is a lecture course it does not cover the entire literature, nor does it cite all authors. The examples used to illustrate principles are often taken from my own work, on which I can sound more convincing.

Whenever possible I have presented a historical introduction. I believe this is necessary for getting the right perspective.

The analytical applications of chromatography are not surveyed as there are excellent treatises available. Besides, I wanted to make the text short and readable.

Lausanne, March 1993   *Michael Lederer*

1 HISTORICAL INTRODUCTION

One of my co-workers who had come to chromatography through a biochemistry course at a good Swiss university expressed his astonishment when he learned that the first paper on chromatography dates from 1903. He told me that from his course he had gained the impression that chromatography had been used all along, not only for the last 90 years or so. It is thus important to summarize the development of the idea of chromatography.

Karl Popper stated that definitions are not very satisfactory, as they are usually based on other definitions. He suggested that, for example, it would be very difficult to give a scientifically exact definition for a 'sand dune' (I suggest that you try it yourself as an exercise), but everybody knows what a sand dune is without giving exact dimensions or a definition of 'sand' etc. So if you ask chromatographers to define 'chromatography' the result is usually poor. The best definition that I have found is that by A. J. P. Martin:

> The technical procedure of analysis by percolation of a fluid through a body of comminuted or porous rigid material, irrespective of the nature of the physico-chemical processes that may lead to the separation of the substances in the apparatus.

Another definition, also by A. J. P. Martin, says 'The essence of the chromatogram is the uniform percolation of a fluid through a column of more or less finely divided substance, which selectively retards by whatever means, certain components of the fluid.'

Please note that both definitions talk about fluids, not liquids. Fluids may be liquids, gases or supercritical fluids.

Considering these definitions, it becomes still more amazing that such systems were 'discovered' so late in the day for chemical analysis, because natural systems are all around us, for example the percolation of rain water through a soil layer.

However, the discovery starts with botanist Michael Tswett, born in 1872. Much historical research has been done in the last twenty years on the life of Tswett by my friend Karel Sakodynskii. His findings make good reading; also he unearthed much photographic material [1].

In chemistry we are often dismayed when a scientific book or paper brings incomplete or irreproducible results. My reading on the life of Tswett gave me the feeling that chemists are still very clear and exact compared with historians.

Two birthdates were claimed, May 14th and May 19th, depending on the source used — school certificates or letters etc. Finally May 14th was decided upon. Tswett was born in Asti (Italy) and his father was Simeon Tswett, a trade representative in Italy of the Russian government. His mother's maiden name was Maria de Dorozza. From this name and the place of birth, some 'historians' deduced that Tswett's mother was Italian. However, she was born in Kütahia (Turkey) and grew up in Russia 'in the family of the brothers Zhemchuzhnikov, cousins of the writer Tolstoi'. The name de Dorozza sounds Italian but it does not seem to be an Italian name. At least, I have not discovered it in any telephone book in Italy.

More jumping to conclusions, as is practised by historians, is not usual even in theoretical chemistry!

Tswett died in 1919 (although some books mention 1920 . . .). Some biographers say he suffered from tuberculosis and others that he 'suffered from a degeneration of the heart'. It seems quite possible that he had both, as at that time few people did not have tuberculosis at some time or another which may not have been completely cured.

How did it come about that a botanist discovered a method which now is used (according to J. Janak) in more than 60% of all analyses that are performed?

The background came to light only recently [2] in the correspondence between Tswett and some of his colleagues. These botanists tried to deal with a rather fundamental question: does a plant have a soul?

To contribute to this question they wanted to study how a plant moves; more precisely they wanted to explain phototropism (heliotropism), the movement of a shoot grown in absence of light which turns towards a light source with simultaneous formation of green pigmentation. According to an elementary university textbook of botany of that time (published in 1894) this phenomenon was well known, and reference to a paper by Wiesner in 1878–80 is given.

Tswett wanted to study the formation of green pigments in this process. At the time the pigments could be separated into various fractions by solvent extraction of leaves. The existence of two chlorophylls and two carotenes was known to Tswett. While filtering extracts through filter paper he observed that the upper border of the paper was more strongly yellow than the bulk extract. Thus a separation had taken place during the process of filtration. Now comes the stroke of genius: Tswett decided that this separative effect could be enhanced by percolating the extract through a column of powdered paper or other solid. This gave many more fractions and Tswett, a good experimentalist, realised the possibilities of such an arrangement and appreciated the underlying dynamic processes.

Tswett published on chromatography from 1903 until about 1910. Some papers were in German and most in Russian. His first paper and another on chlorophylls from 1907 (both in German) have been reprinted [3,9] and should be available in most university libraries; see Figure 1. Some of his 'historians' maintain that Tswett did not understand the importance of his discovery. His papers prove the contrary. Not only did he design apparatus, he examined more than a hundred adsorbents and numerous solvents as eluents.

He distinguished the three modes in which chromatograms can be obtained:
Elution, that is, by washing a narrow adsorbed band with pure solvent.
Displacement, by adding a more adsorbed solvent or solute to desorb the sample.

Figure 1. Left: Book of M.S. Tswett, presented as doctoral thesis. Right: First chromatographic unit and chromatograms [1]

Frontal analysis which is the chromatogram obtained by simply running the mixture to be analysed on a column without washing with pure solvent.

These modes are best explained by the diagrams from an article by A. J. P. Martin published in *Endeavour*; see Figures 2–7 [8].

Last but not least, Tswett invented the term 'chromatography'. He states: 'Like rays of light in the spectrum, different components of the pigment mixture are naturally arranged in the column of calcium carbonate, which allows their qualitative and quantitative determination. The preparation thus obtained I name a chromatogram, and the method proposed — chromatography'.

What happened then? Then, as now, a new idea was unpopular. Like the average referee of scientific journals today, his colleagues did not grasp the fundamentally new idea and started nitpicking. The death sentence was given to chromatography, however, by the greatest organic chemist of his time, R. Willstätter. He had obtained the thesis of Tswett (published in Russian in 1910) and asked one of his co-workers to translate it into German (a handwritten translation was the result of this effort). He then repeated some of the work on adsorbents which were too 'strong' and obtained chemical alterations during chromatography, something which Tswett had also reported. However, it was then concluded that chromatography does not work because it alters the chromatographed substances. Of course, since then the possibility of alterations during chromatography has become well known in many fields and has never deterred anybody.

However at the time this meant no more chromatography ... until 1931. Such generalizations with wrong conclusions are alas not uncommon in science. I shall cite another case. Madame Curie once wanted to purify one gram of radium chloride and tried to recrystallize it by addition of ethanol to an aqueous solution of $RaCl_2$. The ethanol decomposed rapidly under the intense radiation from such an amount of radium. It was thus concluded that one cannot use organic solvents or reagents with radioelements. Thus, no organic solvents were used for many years.

The next chapter in the history of chromatography took place in Heidelberg. One of the best students of Willstätter, Richard Kuhn, had started there a group on natural product chemistry. In 1931 he accepted a young co-worker from his home town of Vienna, Edgar Lederer, and put to him a problem on the carotenoids of egg yolk [4]. Kuhn had taken with him from Munich the German handwritten translation of Tswett's thesis. Edgar Lederer read it and decided to use it on his egg yolk carotenes. He applied chromatography on $CaCO_3$ and obtained excellent results [5]. Thus, after about 20 years of neglect, chromatography scored an important success in natural product chemistry. It has never stopped since. First everybody in Kuhn's laboratory used it, and by 1937 there appeared the first book on

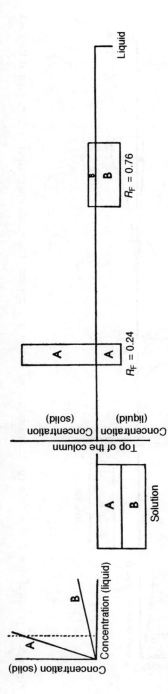

Figure 2. Representation of an ideal chromatogram. Instantaneous equilibrium, no diffusion, and linear adsorption isotherm, the latter shown on the left. On the right is shown the result of putting on the column a solution containing equal amounts of two substances A and B. The bands of A and B stay of unchanged shape

$$R_F = \frac{\text{Movement of band}}{\text{Movement of liquid}} = \text{proportion of solute in liquid phase}$$

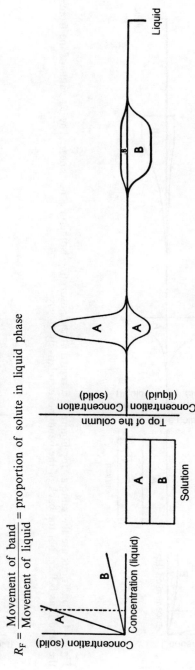

Figure 3. Conditions are the same as for Figure 2, except that equilibrium is not attained instantaneously. The bands assume the shape of the normal curve of error. If measurements are made from the centre of the band, the R_F value remains unchanged. If the bands are originally narrow, their width is proportional to the square root of the distance each has moved down the comlumn

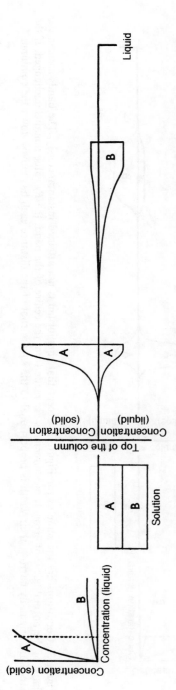

Figure 4. Conditions are the same as for Figure 2, except that the adsorption isotherm is not linear. The R_F value of the front of the band is as before, but the back of the band moves relatively slowly

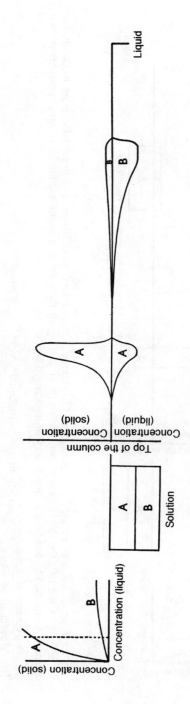

Figure 5. A practical case. None of the conditions stipulated in Figure 2, applies. The bonds are sharper in front than at the back

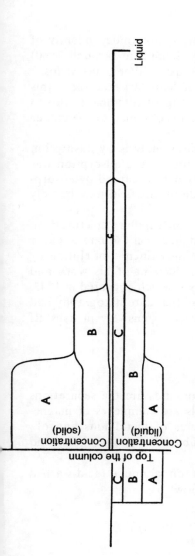

Figure 6. Example of front analysis. When solution is run continuously into the column successive fronts are formed, each step representing a fresh substance. The more strongly adsorbed substances to some extent elute those less strongly adsorbed. This method is due to Tiselius, who has used optical methods to determine concentrations in each step

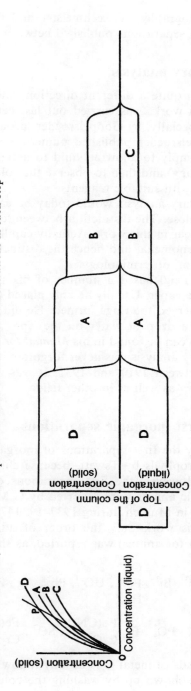

Figure 7. Example of displacement development. The solution is run into the column and is followed by a displacing solution of a strongly adsorbed substance D. The concentrations in the liquid and solid phase are given by the intersections of the line OP with the adsorption isotherms of substances A, B, and C; P is fixed by the concentration of the solute D. Thus for a given concentration of the solute D the distance between the steps are a measure of the amount of each substanct (From *Endeavour*, Vol. 6 No. 21, 1947)

chromatography by Zechmeister and Cholnocky [6] which covered hundreds of separations published between 1931 and 1937.

Capillary analysis

Work in quite a different direction, and seemingly done independently of Tswett's work, was carried out last century by Schönbein (around 1860) and, especially, by Goppelsroeder (around 1900) using filter paper strips.

Goppelsroeder published numerous studies of 'capillary analyses', this being simply to allow a liquid to ascend on a strip of filter paper (up to 50 cm long) and then to observe the coloured bands obtained or the colour formed with suitable reagents.

Capillary analysis would today be called frontal analysis by adsorption on cellulose. The distinction between 'capillary forces' and adsorption was never clear in this work. Actually capillary movement is caused by adsorption phenomena, and hence the distinction made in this field may be only a question of terminology.

Fritz Feigl based a number of his spot tests on capillary 'pictures' on cellulose paper. Usually he only placed one drop on a small square of paper and observed the rings formed. He did also use the principle of elution by placing a drop of water on the 'spot'. A good account of his work and findings can be found in his *Manual for Spot Tests*, which appeared in 1943. Capillary analysis was never forgotten in the way that chromatography had been between 1910 and 1931. It was always used extensively in dyestuff chemistry as well as in other fields.

The first inorganic separations

Actually the first separations of inorganic compounds are the separations of chlorophylls by Tswett, because chlorophylls are complexes of magnesium. However, the first separations of mixtures of metal ions and of inorganic anions were reported by G.M. Schwab and his co-workers, who worked in Munich during 1937–1939 [7].

In this work only the order of adsorption on alumina (acid-washed alumina for anions) was reported, as shown below:

$$As^{3+}, Sb^{3+}, Bi^{3+}, \begin{matrix}Cr^{3+}\\Fe^{3+}\\Hg^{2+}\end{matrix}, UO_2^{2+}, Pb^{2+}, Ag^+, Zn^{2+}, \begin{matrix}Co^{2+}\\Ni^{2+}\\Cd^{2+}\end{matrix}, Tl^+, Mn^{2+}$$

and OH^-, PO_4^{3-}, F^-, $\begin{matrix}Fe(CN)_6^{4-}\\CrO_4^{2-}\end{matrix}$, SO_4^{2-}, $\begin{matrix}Fe(CN)_6^{3-}\\Cr_2O_7^{2-}\end{matrix}$, Cl^-, NO_3^-, MnO_4^-, ClO_4^-, S^{2-}

The bands of metal ions usually follow one on the other (by displacement) and are shown up by washing the column with reagents such as aqueous

H_2S. No 'complete' separations nor quantitative analyses were obtained at that stage. Today one can ask what the purpose was of this work.

I think it was a logical consequence of the extensive use made of chromatography by organic chemists. The standard practical organic chemistry textbook by Gatterman contained a detailed experiment on the chromatographic separation of pigments from leaves. So the question as to what would happen with inorganic ions obviously arose. This was answered by Schwab *et al.* at the time. The 'mechanism' of the adsorption of metal ions is not straightforward, and will be dealt with in the section on inorganic ion exchangers in Chapter 6.

The next stage in the history is best approached by the following discussion on solvent extraction (see Chapter 2).

References

[1] K. I. Sakodynskii, *Michael Tswett, Life and Work*, Carlo Erba Strumentazione, Milan, 62 pp.; *J. Chromatog.* **49** (1970)2; **73**(1972)303; **220**(1981) 1.
[2] I. Hais, *J. Chromatog.* **440** (1988) 509.
[3] R. G. Magee, *Selected Readings in Chromatography*, Pergamon Press Ltd., Oxford (1970), pp. 8–13.
[4] L. S. Ettre and A. Zlatkis, *75 Years of Chromatography — a Historical Dialogue*, Elsevier, Amsterdam (1979), pp. 237–245.
[5] R. Kuhn, A. Winterstein and E. Lederer, *Hoppe-Seyler's Z. Physiol. Chem.* **197** (1931) 141.
[6] L. Zechmeister and L. Cholnoky, *Die Chromatographische Adsorptionsmethode*, Springer, Vienna (1937).
[7] G. M. Schwab and K. Jockers, *Naturwissenschaften* **25** (1937) 44.
G. M. Schwab and G. Dattler, *Angew, Chem.* **50** (1937) 691.
[8] A. J. P. Martin, *Endeavour* **6** (1947) 21.
[9] G. Hesse and H. Weil, *Michael Tswett's erste chromatographische Schrift*, M. Woelm, Eschwege (1954), 37 pp.

2 SOLVENT EXTRACTION AND THE HISTORY OF PARTITION CHROMATOGRAPHY

Solvent extraction methods in inorganic chemistry are not as recent as chromatography. In 1805 Bucholz reported the solubility of uranyl nitrate in diethyl ether.

We shall now consider some of the solvent extraction data which are relevant to the understanding of partition chromatography.

Table 1 from Vogel's *Quantitative Inorganic Analysis* shows the extraction of a number of metal ions from 6 N HCl into ethyl ether. Hydrated cations such as $Al(H_2O)_6^{3+}$ or $Ni(H_2O)_4^{2+}$ are not extracted; nor are quite a number of metal chloro-complexes such as $RhCl_6^{3-}$.

Table 1. Extraction of metal chlorides by ethyl ether from 6N HCl solutions [1]

Element	Per cent extracted at each extraction	Element	Per cent extracted at each extraction
Sb ($SbCl_3$)	6	Fe ($FeCl_3$)	99
Sb ($SbCl_5$)	81	Hg ($HgCl_2$)	0.2
As ($AsCl_3$)	68	Mo ($MoCl_5$)	80–90
Cu	0.05	Te ($TeCl_4$)	34
Ga	97	Tl ($TlCl_3$)	90–95
Ge	40–60	Sn ($SnCl_4$)	17
Au ($AuCl_3$)	95	Sn ($SnCl_2$)	15–30
Ir ($IrCl_4$)	5	Zn	0.2

Al, Bi, Be, Ca, Cd, Cr, Co, Fe^{II}, Pb, Mn, Ni, Os, Pd, rare earths, Rh, Ag, Th, Ti, W, U and Zr are not extracted.

Very good extraction, 95–99% is obtained for Au(III), which is extracted as an ion pair [$H^+ AuCl_4^-$], and also for Fe(III) and Tl(III), which form similar ion pairs [$H^+ MCl_4^-$]. Other strong complexes of metals, e.g. $HgCl_4^{2-}$, are not well extracted. Thus complexation may promote good extraction but need not always do so.

The variation of percentage extraction with the molarity of the acid in the aqueous phase is shown graphically below, in Figures 1–6. Some metals, e.g. Ga(III) and Fe(III), are 99% extracted from 6 N HCl but less completely from higher concentrations of HCl.

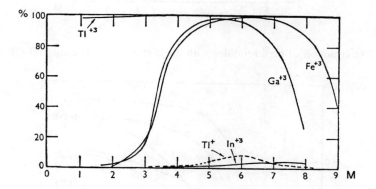

Figure 1. Extraction of metal chlorides with ethyl ether from aqueous HCl [2]

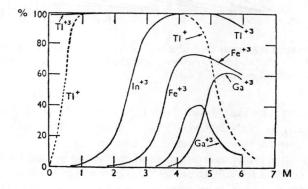

Figure 2. Extraction of metal bromides with ethyl ether from aqueous HBr [2]

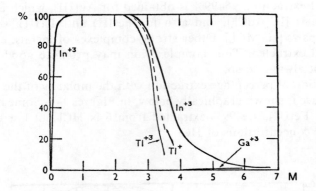

Figure 3. Extraction of metal iodides with ethyl ether from aqueous HI [2]

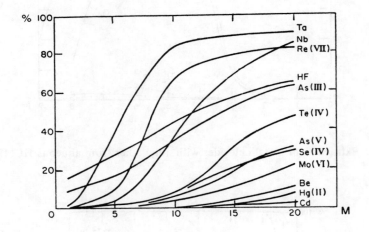

Figure 4. Extraction of metal fluorides with ethyl ether from aqueous HF. Above 20 M HF Mn, Co, Ni and U(VI) extract less than 4%, Zn, Fe(II) and Fe(III) less than 0.3% and Sb(V) less than 0.1% [2]

Uranyl nitrate is extracted well from HNO_3 but still better from solutions of nitrates such as $NaNO_3$, $Ca(NO_3)_2$ or $Al(NO_3)_3$. This 'salting out' is a general effect, and a typical system is shown in Figure 7.

Solvent extraction is an important method in uranium production and purification. A wealth of data for 'salting out' etc. can be found in Volume D2 of Gmelin, which deals entirely with solvent extraction methods for uranium [3].

Solvent extraction is an equilibrium process between two immiscible phases, and the equilibrium constant is usually called the distribution or partition coefficient:

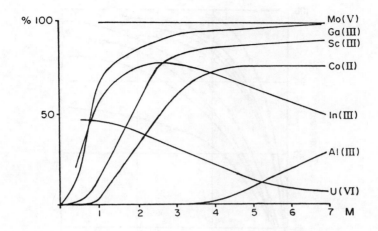

Figure 5. Extraction of metal ions (initially 0.1 M) with ethyl ether from a solution which is 0.5 M in HCl with additions of ammonium thiocyanate (0–7 M) [2]

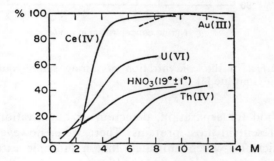

Figure 6. Extraction of metal nitrates with ethyl ether from aqueous HNO_3 [2]

$$\frac{\text{Concentration of solute in solvent A}}{\text{Concentration of solute in solvent B}} = \frac{C_A}{C_B} = K$$

when equal volumes are shaken till equilibrium is attained.

To continue with the history of chromatography. A.J.P. Martin was born in 1910 in London, the son of a medical doctor. He first wanted to become an engineer, but after four years as an engineering student he suddenly decided that biochemistry was more interesting and completed the biochemistry course in two years. After some work on vitamins he joined the Wool Industries Research Organisation in 1938 where, together with R. L. M. Synge, he started work on wool proteins. At the time there was no

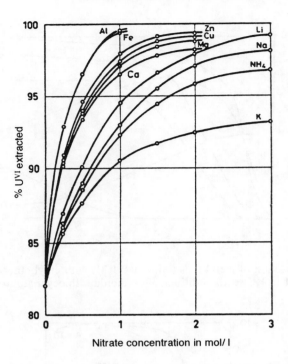

Figure 7. The effect of the concentration of nitrate salting out agents on the extraction of uranyl nitrate [3]

analytical method for separation, detection or quantitation of the 20–25 amino acids present in most proteins. There were, however, data on the partition coefficients, which indicated that a multiple extraction system should be able to solve at least some difficult analytical problems. Martin and Synge therefore built multiple extraction set-ups and published results on some shorter peptides.

In his memoirs Martin relates [4]:

> We were eventually successful in using this machine to separate, and measure fairly accurately, the monoamino, monocarboxylic acids in wool. It was a fiendish piece of apparatus, we had to sit by it for a week for one separation; it had 39 theoretical plates and filled the room with chloroform vapour. We used to watch it in 4-hour shifts. We had constantly to adjust small silver baffles to keep the apparatus working properly. One of the effects of four hours of chloroform intoxication was that when our partner arrived to take the next shift he was invariably sworn at by the one who had been watching the machine. Another curious effect of the chloroform was that when I went into the fresh air, it smelt peculiar. This was my first experience of the interesting phenomenon of negative smell and may have been partly responsible for my current interest in the physiology of

the sensation of smell. (*Reproduced by permission of Elsevier Science Publishers BV*).

Solvent–solvent extraction machines were then improved both by Martin and Synge and by L.C. Craig. A modern apparatus uses unit cells, as shown in Figure 8. These cells are usually assembled in automatic machines, permitting different equilibration times, settling and decantation times etc. Usually machines for 50 or 100 units are made and higher numbers of equilibria are obtained by recycling.

Figure 9 show a typical 'Craig machine' complete with fraction collectors and encased in a housing with ventilation [5].

Figures 10 and 11 show some results that can be obtained with such machines. Although they are slow and have relatively few equilibria (usually 100 and, at most, 1000 with recycling) they have the advantage of quite large volumes (usually 25 or 50 ml) of each phase and the possibility of calculating the theoretical distributions with good accuracy.

Solvent extraction batteries are used today in the industrial separation of rare earth elements, by using 50 extractors with a gradient extraction system which concentrates each rare earth element in one or a few extractors. There are several advantages of gradient solvent extraction compared with the previously used preparative scale ion exchange columns — higher capacity,

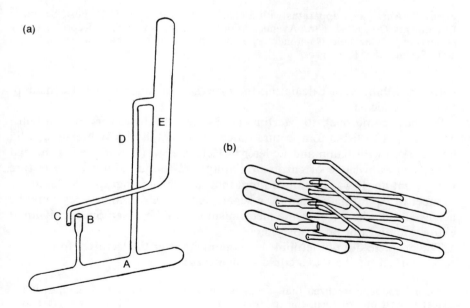

Figure 8. (a) A unit cell of a Craig machine [2]. (b). Arrangement of the unit cells in a Craig machine [2]

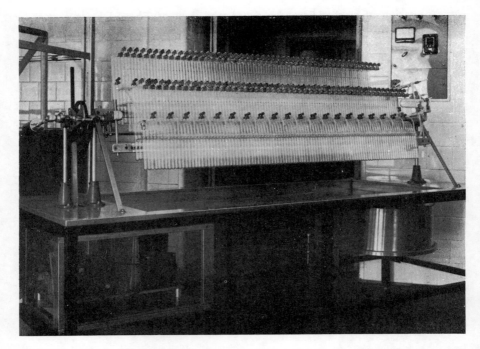

Figure 9. Automated apparatus with 200 cells, constructed by H.O. Post, Scientific Instruments Co., 6957, 63rd Avenue, Middle Village, NY, USA. (Reproduced by permission of Consiglio Nazionale delle Ricerche, Ufficio Pubblicazioni e Informazioni Scientifiche, Rome)

better possibilities of calculating the process and much reduced consumption of chemicals.

We now come back to Martin and Synge and their work on proteins. Martin went to listen to a lecture on chromatography by Winterstein (who had worked with Kuhn and Lederer in Heidelberg). He told me that he had then little knowledge of chromatography and found Winterstein's lecture wildy exciting. It was obvious to him that in a chromatographic column the number of equilibria needed for a separation of many amino acids could be achieved, only he felt that the mechanism should be based on liquid–liquid extraction and not on adsorption.

The problem did not let him sleep the night after the lecture. However he soon worked out how this could be achieved:

> Then I suddenly realized that it was not necessary to move both liquids; if I just moved one of them the required conditions were fulfilled. I was able to devise a suitable apparatus the very next day, and a modification of this eventually became the partition chromatograph with which we are now familiar. Synge and I took silica gel intended as a drying agent from a balance case,

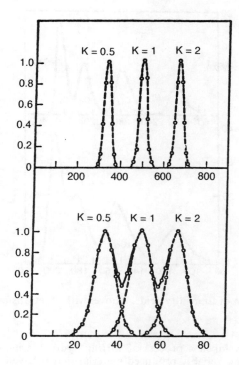

Figure 10. Separation of a three-component mixture. Above: with 1000 transfers; below: with 100 transfers [6]

ground it up, sieved it and added water to it. We found that we could add almost its own weight of water to the gel before it become noticeably wet. We put this mixture of silica gel and water into a column, put the acetylamino acids on to the top and poured chloroform down the column. We wondered how we should know where the amino acids were in the column and when to expect them to emerge at the bottom of the tube. By the end of the first day there was no sign of them. To find out what was happening in the column we added methyl orange to the liquid on the silica gel and thus were able to see the acetylamino acids passing down the column as a red band. One foot of tubing in this apparatus could do substantially better separations than all the machinery we had constructed until then.

Normal chloroform contains about 1% of ethyl alcohol as a stabilizer. The first experiment we did, with chloroform straight out of a bottle in the laboratory, gave the results we expected, and we separated acetylproline and acetylleucine. We next used carefully distilled chloroform and were surprised to find that the amino acids did not move from the top of the column. The reason for this, of course, was increased adsorption due to the absence of ethyl alcohol. When we added alcohol to the chloroform our bands could move down the column again. But it was difficult to produce a satisfactory colour change; we needed large amounts of acids with our original silica gel–methyl orange system. So we experimented with different ways of making precipitated silica,

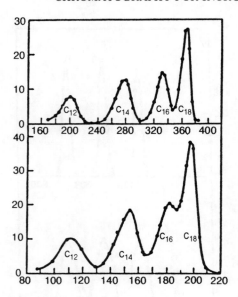

Figure 11. Separation of four fatty acids. Above: with 400 transfers; below: with 200 transfers [6]

and eventually developed a process of stirring hydrochloric acid into sodium silicate. This process reliably produced material that behaved as we wished in the columns. But this work was more magic than science; we never understood in detail what we were doing. Later we changed the indicator; at one time we used pelargonin which we extracted ourselves from various flowers.

This work was eventually published in 1941. In this paper we noted that the mobile phase could just as well be a gas as a liquid, We also predicted that with such a system, very refined separations of various kinds of compounds would be possible. Although this paper was widely read by chemists in different fields, no one thought this prediction worth testing experimentally.

In spite of all our efforts we could separate only the monoamino, monocarboxylic acids. We could not make the system work for the dicarboxylic or basic amino acids. So we looked for materials other than silica to hold the water and our first choice was paper. I had seen paper chromatograms of dyes and was familiar with the uptake of water by cellulose, so paper was an obvious choice. Dr. A.H. Gordon, who was now working with us at Leeds, looked through *Beilstein's Handbuch der Organischen Chemie* to find a colour reaction that would reveal our amino acids on the paper; he found ninhydrin which proved admirable for our purpose. Our first paper chromatograms were circles of paper in a Petri dish containing water and water-saturated butanol fed by capillarity to the centre by a tail on which a drop of amino acid solution had been placed. When the butanol reached the edge, the paper was dried and sprayed with ninhydrin in dry butanol. Later, we used strips of paper in test-tubes and more suitable containers — boxes in which the air was kept saturated with water — with troughs containing the mobile solvent into which the tops of the strips could dip. Several boxes were needed since it was characteristic of the method that though it was not particularly quick, very little work was needed to run many strips simultaneously. An important step was running the chromatogram in two dimensions. The first solvent spread the amino acids in a line near one end of the paper from a spot near the corner; then, after drying, we turned the paper through a right angle and spread the line of spots into a two-dimensional pattern by using a different solvent [3].

I would like to mention here that the first two-directional separation that I did was with electrophoresis in one direction in an acetate buffer and paper chromatography in the other. The track for the electrophoresis was isolated from the rest of the column by saturating the paper with paraffin wax on either side. But chromatography turned out to be more satisfactory than electrophoresis at that time, so I did not work with electrophoresis again until 1944.

Our next problem was to deal with the curious fact that in some, but not other, solvents the purple amino acid spots had a pink "beard" underneath them; and as they ran further down the paper the purple colour showed less and the pink more. The purple-coloured spots of leucine and phenylalanine, for example, had almost vanished before they got to the bottom of the paper, leaving only a faint pink blob. This unsatisfactory colour change was particularly marked with papers on which the chromatogram had been run in two directions: in phenol in one direction and in collidine in the other. This two-directional system was convenient because we could distinguish between the acidic, basic and neutral compounds. However, when the separation was run in an atmosphere of ammonia so as to increase the pH, a further problem was observed: the paper became covered with black spots.

Eventually we found the cause of these problems. We identified the cause of the black spots on the paper as copper from the fans used for drying the papers in the laboratory; these fans had a badly sparking commutator which filled the room with copper. The large amounts of copper in the Leeds atmosphere also contributed to the copper on our papers. The black colouration was due to the catalytic oxidation of phenol by copper in the presence of ammonia, and the pink beards were caused by a copper complex of the amino acid that formed on the paper. The formation of these complexes could be suppressed by including a complexing agent for copper — cyanide for example — in the coal gas we put into the atmosphere in the box [4]. (*Reproduced by permission of Elsevier Science Publishers BV*).

The first paper submitted in this new form of chromatography referred to liquid–liquid chromatography in the title to distinguish it from liquid–solid chromatography. The editor of the *Biochemical Journal*, however, did not like this term and altered it to 'liquid–liquid partition chromatography' This does not improve the term, as partition can equally refer to liquid–solid. The term 'partition chromatography' has however been adopted by chromatographers as meaning liquid–liquid systems.

Almost simultaneously with Martin and Synge, Boldingh [7] used moderately vulcanized Hevea rubber to support liquid benzene as the stationary phase in separations of higher fatty acids by eluting with methanol, i.e. a more polar solvent. When the non-polar phase is stationary the term 'reversed phase' partition chromatography is now used. There is of course no reason why this system is called reversed and not the one with the polar stationary phase.

The term 'paper chromatography' has nowadays the meaning of liquid–liquid chromatography on paper. One can also effect adsorption or ion exchange chromatography on paper or even reversed phase chromatography (on paraffin impregnated paper), but such techniques are usually referred to as adsorption paper chromatography etc.

Now for a brief account of the principles of paper chromatography.

THE R_F VALUE

The movement of spots on paper chromatograms is usually recorded in terms of R_F values, where

$$R_F = \frac{\text{distance travelled by spot (centre)}}{\text{distance travelled by the liquid front}}$$

The R_F is used in a paper on column chromatography by LeRosen [8] and is meant to be the *rate*, with F indicating that it is the rate relative to the liquid front. Recently this was also called 'retention factor', which is obviously wrong, as it does not increase but decreases with retention.

In a way, as we shall see below, the term R_F was not a fortunate choice, as it is not directly related to the equilibrium of the system. However, it became popular because it is easy to measure (in mm) with a ruler. Accuracy of measurement and experimental conditions do not permit an accuracy better than about ± 0.02; R_F values to three decimal places require much optimism and fantasy.

THE R_M VALUE

In most (partition) paper chromatographic systems adsorption and ion exchange are negligible, and thus the R_F value reflects the liquid–liquid extraction behaviour of the chromatographed substances. However, the extraction or partition coefficient is proportional not to the R_F value but to $(1/R_F - 1)$.

The partition coefficient can be calculated by the equation

$$\alpha = \left(\frac{1}{R_F} - 1\right)\frac{A_L}{A_S}$$

where A_L/A_S is the ratio of the mobile phase to the stationary phase. This is usually not determined (too cumbersome); for most purposes it is derived from a partition coefficient determined in equilibrium experiments or considered as a 'system constant', as for all substances chromatographed in the same system it is identical.

At the Biochemical Symposium on Partition Chromatography in 1948, Martin pointed out that [9]:

to a first approximation $\Delta\mu_A$ may be regarded as made up of

$$d\Delta\mu_{CH_2} + e\Delta\mu_{COO^-} + f\Delta\mu_{NH_3^+} + g\Delta\mu_{OH} + \ldots, \text{etc.}$$

the sum of the potential differences of the various groups of which the molecule A is composed. That is to say, to a first approximation the free energy required to transport a given group, e.g. CH_2, from one solvent to another is independent of the rest of the molecule. Thus all isomers containing the same groups (note that the degree of ionization, etc., must not be changed) would be expected to have the same partition coefficient.

Now, if we consider the partition coefficients α_A and α_B of two substances A and B which differ in that B contains, in addition to those contained in A, a group X, we have,

$$\ln \alpha_A = \frac{\Delta\mu_A}{RT}, \quad \ln \alpha_B = \frac{\Delta\mu_B}{RT} + \frac{\Delta\mu_X}{RT}, \quad \ln\left(\frac{\alpha_B}{\alpha_A}\right) = \frac{\Delta\mu_X}{RT}$$

Thus the addition of a group X changes the partition coefficient by a given factor depending on the nature of the group, and on the pair of phases employed, *but not on the rest of the molecule.*

This is a prediction contrary to usual expectation. It is usually felt that the formation of a derivative of greatly increased molecular weight will 'swamp' any differences that exist and will render separation more difficult. This, however, is not to be expected if such a derivative be chosen that the same pair of phases can be employed while still maintaining convenient values for the partition coefficients.

Let us apply this rule to amino-acids and peptides. On the formation of a dipeptide molecule from two amino-acid molecules, or a tripeptide from an amino-acid and a dipeptide, one –CONH– group is created and one COO^- and one NH_3^+ are destroyed. Let the amino-acids be $A_{COO^-}^{NH_3^+}$ and $B_{COO^-}^{NH_3^+}$ and the peptide $NH_3^+.A.CO.NH.B.COO^-$, and let the partition coefficients be α_A, α_B and α_{AB} respectively,

$$RT \ln \alpha_A = \Delta\mu_{NH_3^+} + \Delta\mu_A + \Delta\mu_{COO^-}$$
$$RT \ln \alpha_B = \Delta\mu_{NH_3^+} + \Delta\mu_B + \Delta\mu_{COO^-}$$
$$RT \ln \alpha_{AB} = \Delta\mu_{NH_3^+} + \Delta\mu_A + \Delta\mu_{CONH} + \Delta\mu_B + \Delta\mu_{COO^-}$$

$$RT \ln\left(\frac{\alpha_A \alpha_B}{\alpha_{AB}}\right) = \Delta\mu_{NH_3^+} + \Delta\mu_A + \Delta\mu_{COO^-} - \Delta\mu_{CONH}$$

i.e. the product of the partition coefficients of the constituent amino-acids divided by the partition coefficient of the dipeptide is a constant for any given phase pair.

These considerations are fundamental for understanding how different molecules will be separated, but unless we have data on partition coefficients we can not apply them.

Bate-Smith and Westall [10] suggested a simple way out of this. They proposed the R_M value:

$$R_M = \log\left(\frac{1}{R_F} - 1\right)$$

This R_M value is proportional to the free energy; within a solvent system all the factors R, T, A_L, A_S are the same for all compounds, and this permits calculations on the relationship between the structure of a compound and its R_F value. For the conversion of R_F values to R_M values see Table 2.

The equation which is now used is

$$R_M = a\Delta R_{M_{CH_2}} + b\Delta R_{M_{COOH}} + c\Delta R_{M_{NH_2}} \ldots + K$$

where K is referred to as the solvent constant.

Table 2. Conversion of R_F to R_M values; $R_M = \log\left(\frac{1}{R_F} - 1\right)$

R_F	R_M	R_F	R_M	R_F	R_M	R_F	R_M
0.01	1.999	6	454	2	34	7	524
2	690	7	432	3	52	8	550
3	510	8	411	4	69	9	575
4	380	9	389	5	87	0.80	−0.602
5	279	0.30	0.368	6	−0.105	1	631
6	195	1	347	7	122	2	660
7	128	2	327	8	140	3	690
8	061	3	307	9	158	4	721
9	005	4	288	0.60	−0.176	5	754
0.10	0.95	5	269	1	194	6	791
11	91	6	250	2	212	7	827
2	87	7	231	3	234	8	866
3	83	8	212	4	250	9	910
4	79	9	194	5	269	0.90	−0.955
5	75	0.40	0.176	6	288	1	−1.004
6	72	1	158	7	308	2	061
7	69	2	140	8	327	3	125
8	66	3	122	9	348	4	194
9	63	4	105	0.70	−0.368	5	276
0.20	0.600	5	087	1	389	6	377
1	575	6	070	2	410	7	509
2	549	7	052	3	433	8	699
3	528	8	035	4	456	.9	999
4	501	9	017	5	477		
5	477	0.50	0.000	6	500		
		1	−0.017				

A typical example of structure – R_F correlations using the R_M concept is shown below.

Table 3. Resolution of the R_M Values into Partial Constants [11]

Solvent system	Amyl alcohol 1 5 N HCOOH 1	Ethyl alcohol 80 NH_3 (d = 0.088) 4 H_2O 36	Phenol saturated with H_2O Cupron 0.1%
Paper	Whatman No.1	Whatman No. 54	Whatman No. 1

	Constants expressed by the R_M values		
Constant for the paper and system (G_P)	−0.97	−0.43	−0.57
Group constant G — Every C atom	−0.12	−0.08	−0.27
Branching of the chain	−0.25	−0.05	−0.07
Closing of the benzene ring			−0.61
Benzene ring			−0.01
Primary OH (alcohol)	−0.73	−0.20	−0.36
Secondary OH (alcohol)	−0.50	−0.13	−0.38
Tertiary OH (alcohol)	−0.58		
OH (Phenolic)			−0.91
α-CO (Ketone)	−0.39		
COOH	−0.63	−0.56	−1.07
α-NH_2	−1.65	−0.24	−0.20
δ-NH_2			−0.89
ε-NH_2			−0.96
Imidazole group			−0.25
–S–			−0.02
–S–S–			−0.04

	Examples R_F values					
	found	calc.	found	calc.	found	calc.
Propionic acid			0.56	0.56		
Lactic acid	0.62	0.61	0.49	0.49		
Pyruvic acid	0.65	0.67				
Malonic acid	0.53	0.54	0.26	0.26		
Butyric acid			0.64	0.61		
α, β-Dihydroxybutyric acid	0.40	0.40				
α, γ-Dihydroxybutyric acid	0.29	0.28				
Succinic acid	0.61	0.61	0.29	0.30		
Malic acid	0.32	0.33	0.25	0.24		
Tartaric acid	0.14	0.13	0.19	0.19		
Aspartic acid	0.03	0.03	0.18	0.20	0.17	0.17
Alanine			0.43	0.43	0.59	0.56
Serine			0.32	0.32	0.36	0.36
Lysine					0.48	0.48
Cystine					0.29	0.29
Methionine					0.81	0.81
Tyrosine					0.62	0.62
Histidine					0.69	0.70

usually the solvent constant is calculated from one or more substances whose R_F was measured. The calculated R_F values of the other substances are usually within ± 0.02 of the measured ones. There are some splendid correlations for smaller peptides, as shown in Table 4.

THE THEORETICAL PLATE CONCEPT

In fractional distillation the HETP (height equivalent of a theoretical plate) is a useful criterion to calculate the efficiency of the distillation. Similarly, this concept was tried in chromatography, going back to the liquid–liquid extraction systems.

It can be assumed that equilibrium is established in an extraction tube (if shaken to equilibrium). If we now consider a 1:1 distribution this will give us distribution curves depending on the number of tubes employed, as shown in Figure 12. The effect of the number of tubes is seen still better in Figures 10 and 11, where the same separation is shown after 200 and 400 tubes.

The mathematics of such a binomial expansion are of course not simple, but a good approximation formula is generally used

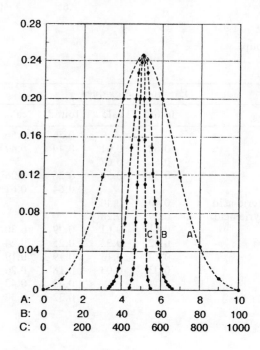

Figure 12. Distribution curves with $\alpha = 1$, $R_F = 0.5$ with 10, 100 and 1000 distribution steps [6]

SOLVENT EXTRACTION AND THE HISTORY OF PARTITION 25

$$N = 16 \left(\frac{l}{w}\right)^2$$

where N is the number of theoretical plates, l is the distance moved by a zone and w is the zone width.

In paper chromatography the visible border of a spot is a good approximation for this. Paper chromatograms with 50 cm development give 800–2500 plates, Thin layer chromatograms developed for about 10 cm up to 4000–5000 plates. A capillary gas chromatogram will have around 100 000 plates. Capillary HPLC and capillary zone electrophoresis can reach several hundred thousand plates or even a million.

The utility of this whole concept is still questioned. Some authors like to talk rather of spot capacity of a chromatogram, i.e. how many completely separated zones can be formed.

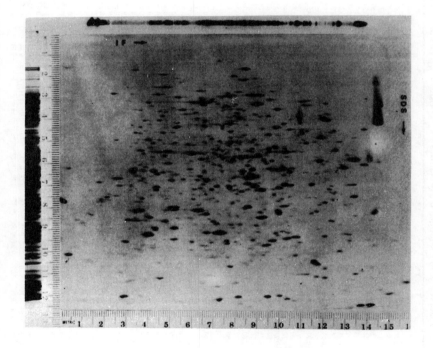

Figure 13. Separation of *E. coli* proteins. *E. coli* was labelled with ^{14}C amino acids *in vivo*. The cells were lysed by sonication, treated with DNase and RNase and dissolved in 9.5 M urea, 2% Nonidet P-40, 2% Ampholine, pH 5–7, and 5% 2-mercaptoethanol. 25 µl of sample containing 180 000 c.p.m. and approximately 10 µg of protein was loaded in the gel in the first dimension (IEF). The gel in the second dimension (SDS) was a 9.25–14.4% exponential acrylamide gradient. At an exposure of 825 h, it is possible to count 1000 spots on the original autoradiogram [13]. (Reproduced by permission of Elsevier Biomedical Press)

Table 4. Comparison of Observed and Calculated R_F Values of Peptides [12]. (Reproduced by permission of Elsevier)

Peptide	A* Obs.	A* Theory	Δ	B* Obs.	B* Theory	Δ	E* Obs.	E* Theory	Δ	F* Obs.	F* Theory	Δ	G* Obs.	G* Theory	Δ	H* Obs.	H* Theory	Δ	K* Obs.	K* Theory	Δ	M* Obs.	M* Theory	Δ	O* Obs.	O* Theory	Δ
Glycylglycine	0.32	0.37	0.05	0.22	0.22	0.00	0.37	0.37	0.00	0.61	0.56	-0.05	0.15	0.11	-0.04	0.18	0.22	0.04	0.19	0.20	0.01	0.46	0.34	-0.12	0.41	0.33	-0.03
Glycylsarosine	0.42	0.43	0.01	0.17	0.20	0.04	0.68	0.75	0.07	0.80	0.81	0.01	0.44	0.06	-0.38	0.33	0.00	0.18	0.29	0.11	0.52	0.49	-0.03	0.43	0.56	0.13
Alanylglycine	0.47	0.47	0.00	0.23	0.32	0.09	0.55	0.53	-0.02	0.70	0.66	-0.04	0.25	0.22	-0.08	0.33	0.31	-0.02	0.27	0.28	0.01	0.53	0.48	-0.10	0.51	0.50	0.01
Leucylglycine	0.75	0.75	0.00	0.54	0.65	0.11	0.68	0.76	0.08	0.86	0.85	-0.01	0.56	0.59	0.03	0.61	0.51	-0.10	0.60	0.56	-0.04	0.75	0.69	-0.06	0.74	0.74	0.00
Glycylalanine	0.47	0.47	0.00	0.28	0.32	0.04	0.52	0.53	0.01	0.72	0.66	-0.06	0.28	0.22	—	0.30	0.31	0.01	0.27	0.28	0.01	0.56	0.48	-0.08	0.72	0.74	0.00
Sarcosylalanine	0.53	0.53	0.00	0.28	0.33	0.08	0.83	0.85	0.02	0.82	0.37	0.05	—	—	—	0.43	0.02	0.41	0.39	0.28	0.59	0.67	0.54	0.68	0.50	0.50	0.00
Alanylalanine	0.58	0.57	-0.01	0.38	0.47	0.09	0.66	0.68	0.02	0.77	0.74	-0.03	0.39	0.40	0.01	0.40	0.41	0.01	0.33	0.37	0.04	0.64	0.69	0.05	0.58	0.62	0.04
α-Glutamylalanine	0.53	0.48	-0.05	0.25	0.25	0.00	0.32	0.22	-0.10	0.48	0.40	-0.08	0.03	0.03	0.00	0.54	0.23	0.02	0.08	0.09	0.01	0.65	0.65	0.09	0.58	0.35	0.07
Glycylserine	0.27	0.34	0.07	0.15	0.19	0.04	0.29	0.29	0.00	0.53	0.51	-0.02	0.10	0.05	-0.03	0.19	0.24	-0.05	0.17	0.32	0.15	0.44	0.37	-0.07	0.41	0.39	-0.02
Glycyl-α-amino-n-butyric acid	0.59	0.66	0.07	0.36	0.38	0.02	0.63	0.69	0.06	0.78	0.79	0.01	0.39	0.39	0.00	0.53	0.41	-0.11	0.31	0.33	0.02	0.62	0.58	-0.04	0.58	0.54	-0.04
Alanyl-α-amino-n-butyric acid	0.70	0.74	0.04	0.48	0.54	0.06	0.75	0.81	0.06	0.84	0.85	0.01	0.53	0.60	0.07	0.53	0.52	-0.01	0.37	0.44	0.07	0.69	0.74	0.05	0.65	0.66	0.01
Glycylthreonine	0.35	0.40	0.05	0.24	0.20	-0.04	0.41	0.45	-0.04	0.66	0.60	-0.06	0.16	0.16	0.00	0.24	0.33	0.09	0.21	0.53	0.32	0.50	0.51	0.01	0.48	0.47	-0.01
Glycylvaline	0.63	0.70	0.07	0.43	0.49	0.06	0.64	0.73	0.09	0.82	0.82	0.00	0.45	0.49	0.04	0.44	0.48	0.04	0.43	0.43	0.00	0.67	0.67	-0.01	0.58	0.67	0.09
Glycylnorvaline	0.64	0.77	0.13	0.49	0.53	0.04	0.65	0.77	0.12	0.84	0.84	0.00	0.51	0.52	0.00	0.47	0.51	0.04	0.45	0.44	-0.01	0.65	0.64	0.64	0.00
Alanylnorvaline	0.78	0.84	0.06	0.59	0.69	0.10	0.80	0.80	0.00	0.87	0.89	0.02	0.63	0.72	0.08	0.59	0.62	0.03	0.49	0.56	0.06	0.72	0.79	0.07	0.71	0.74	0.03
Glycylleucine	0.73	0.75	0.02	0.61	0.65	0.04	0.76	0.76	0.00	0.86	0.85	-0.01	0.58	0.63	0.01	0.55	0.51	-0.04	0.54	0.55	0.01	0.70	0.69	-0.01	0.72	0.74	0.02
α-Glutamylleucine	0.70	0.76	0.06	0.64	0.65	0.01	0.52	0.43	-0.09	0.63	0.67	0.04	0.50	0.22	0.02	0.68	0.41	-0.27	0.20	0.25	0.05	0.79	0.79	0.06	0.50	0.61	0.11
Glycylisoleucine	0.74	0.80	0.06	0.62	0.63	0.01	0.71	0.79	0.01	0.86	0.86	0.04	0.00	0.61	0.06	0.57	0.55	0.50	0.52	0.03	0.70	0.72	0.02	0.70	0.73	0.03
Glycylnorleucine	0.72	0.85	0.13	0.63	0.67	0.04	0.73	0.81	0.08	0.85	0.88	0.03	0.59	0.64	0.05	0.57	0.58	0.01	0.58	0.57	-0.01	0.70	0.62	-0.01	0.74	0.72	-0.02
Glycyl-α-amine-n-caprylic acid	0.86	0.93	0.07	0.71	0.80	0.09	0.77	0.86	0.09	0.89	0.90	0.01	0.65	0.79	0.14	0.59	0.67	0.08	0.66	0.75	0.09	0.71	0.75	-0.04	0.81	0.87	0.06
Sarcosylphenylalaninle	0.66	0.78	0.12	0.50	0.59	0.09	0.80	0.96	0.16	0.81	0.96	0.15	0.77	0.93	0.16	0.52	0.60	0.08	0.53	0.70	0.17	0.67	0.75	0.08	0.64	0.85	0.21
Alanylphenylalanine	0.76	0.80	0.04	0.62	0.72	0.10	0.76	0.90	0.14	0.87	0.92	0.05	0.62	0.82	0.20	0.56	0.58	0.02	0.56	0.69	0.13	0.73	0.73	0.06	0.75	0.81	0.06
α-Glutamylphenylanlanine	0.73	0.74	0.01	0.62	0.57	-0.05	0.51	0.53	0.02	0.64	0.72	0.08	0.28	0.26	0.06	0.35	0.37	0.02	0.20	0.27	0.07	0.74	0.74	0.06	0.43	0.59	0.16
L-Cystinyl-L-cyatine	0.38	0.01	-0.37	0.02	0.02	0.00	0.46	0.06	-0.40	0.68	0.34	-0.32	0.20	0.01	-0.19	0.32	0.02	0.32	0.42	0.14	-0.28	0.53	0.17	-0.36	0.56	0.08	-0.48
L-Cystinyl-D-cystine	0.08	0.01	-0.07	0.11	0.02	-0.09	0.12	0.06	-0.06	0.36	0.34	-0.02	0.38	0.01	-0.02	0.01	0.02	0.01	0.12	0.14	0.02	0.20	0.17	-0.03	0.31	0.08	-0.23
Glycylmethionine	0.64	0.66	0.02	0.48	0.48	0.00	0.71	0.75	0.04	0.81	0.85	0.04	0.45	0.51	0.06	0.41	0.46	0.03	0.41	0.44	0.03	0.59	0.62	0.03	0.65	0.65	0.01
Glycyltryptophan	0.59	0.55	-0.04	0.51	0.47	-0.04	0.71	0.67	-0.04	0.84	0.79	-0.05	0.49	0.45	-0.04	0.35	0.30	-0.05	0.50	0.56	0.06	0.57	0.40	-0.17	0.64	0.64	0.00
α-Aspartylhistidine	0.12	0.10	-0.02	0.05	0.07	0.04	0.38	0.15	-0.13	0.45	0.44	-0.01	0.61	0.05	0.05	0.04	0.08	-0.05	0.06	0.08	0.02	0.27	0.34	0.07	0.23	0.23	0.00
Glycylaine-H$_2$SO$_4$	0.12	0.20	0.08	0.05	0.09	—	0.51	0.51	0.16	0.83	0.90	0.07	0.51	0.38	-0.13	0.04	0.08	0.01	0.13	0.12	0.14	0.16	0.02	0.41	0.70	-0.05
Glycylaspartic acid	0.16	0.28	0.12	0.12	0.19	0.07	0.35	0.07	-0.28	0.56	0.23	-0.33	0.11	0.07	—	0.12	0.14	0.02	0.12	0.06	-0.06	0.32	0.37	0.05	0.33	0.36	-0.05
Glycylglutamic acid	0.34	0.38	0.04	0.25	0.31	0.07	0.35	0.13	-0.04	0.29	0.31	0.02	0.03	-0.01	0.02	0.07	0.16	0.43	0.47	0.04	0.33	0.24	0.19	0.25	-0.14
α-Glutamylglutamic acid	0.41	0.39	-0.02	0.25	0.20	-0.05	0.07	0.03	-0.04	0.10	0.04	0.04	0.00	0.04	-0.01	-0.04	0.11	0.03	0.60	0.10	0.15	-0.03
Glycyltyrosine	0.50	0.55	0.05	0.25	0.41	0.14	0.56	0.56	0.00	0.74	0.69	-0.05	0.27	0.25	-0.02	0.36	0.36	-0.04	0.39	0.39	0.03	0.54	0.54	-0.05	0.56	0.15	0.04
Glycylglycylglycine	0.24	0.30	0.06	0.15	0.14	-0.01	0.44	0.40	-0.04	0.72	0.63	-0.09	0.22	0.12	-0.10	0.18	0.21	0.03	0.21	0.17	-0.04	0.40	0.25	-0.15	0.34	0.34	0.00
Alanylglycylglycine	0.39	0.39	0.00	0.24	0.24	0.00	0.62	0.56	-0.06	0.79	0.72	-0.07	0.34	0.25	-0.09	0.28	0.29	0.03	0.27	0.24	-0.03	0.45	0.50	-0.04	0.45	0.46	0.01
Leucylglycylglycine	0.66	0.69	0.03	0.50	0.55	0.05	0.81	0.78	-0.03	0.87	0.89	0.02	0.59	0.64	0.05	0.54	0.49	-0.10	0.54	0.51	-0.03	0.74	0.68	-0.06	0.71	0.72	0.01
Glycylalanylglycine	0.35	0.39	0.04	0.50	0.24	0.00	0.58	0.56	-0.02	0.78	0.72	-0.06	0.25	0.25	0.00	0.28	0.29	-0.04	0.23	0.24	0.01	0.52	0.50	-0.02	0.48	0.46	-0.02
Glycylleucylglycine	0.65	0.69	0.04	0.58	0.55	-0.03	0.76	0.78	0.03	0.88	0.89	0.01	0.61	0.64	0.03	0.57	0.49	-0.08	0.55	0.51	-0.04	0.72	0.68	-0.04	0.73	0.71	-0.02
Glycylphenylalanylglycine	0.63	0.66	0.03	0.47	0.47	0.00	0.76	0.84	-0.08	0.84	0.91	0.07	0.59	0.69	0.07	0.44	0.45	0.01	0.49	0.54	0.05	0.65	0.61	-0.04	0.66	0.70	0.04
Glycylglycylalanine	0.38	0.39	0.01	0.24	0.24	0.00	0.56	0.56	0.00	0.81	0.72	-0.09	0.25	0.36	0.11	0.24	0.29	0.00	0.27	0.24	-0.03	0.52	0.50	-0.02	0.47	0.46	-0.01
Glycylglycylleucine	0.65	0.69	0.04	0.54	0.55	0.01	0.82	0.78	-0.04	0.86	0.89	0.03	0.67	0.64	-0.03	0.55	0.49	0.00	0.51	0.51	0.00	0.72	0.68	-0.04	0.68	0.71	0.03
Glycylglycylphenylalanine	0.65	0.66	0.01	0.49	0.47	-0.02	0.82	0.84	0.02	0.88	0.91	0.03	0.55	0.58	0.03	0.42	0.45	0.11	0.50	0.54	0.00	0.65	0.61	-0.04	0.64	0.70	0.06
Glycylglycylcystine	0.07	0.04	-0.03	0.08	0.07	0.05	0.33	0.18	-0.02	0.52	0.34	-0.01	0.05	0.04	-0.01	0.03	0.07	0.03	0.05	0.05	0.00	0.15	0.23	0.08	0.22	0.16	-0.06
Alanylalanylcystine	0.16	0.09	-0.07	0.12	0.13	0.05	0.33	0.18	-0.02	0.65	0.71	0.06	0.24	0.21	0.06	0.09	0.16	0.07	0.09	0.11	0.02	0.29	0.56	0.27	0.34	0.35	0.01
Glycylglycylglycylglycine	0.18	0.23	0.05	0.08	0.10	-0.02	0.58	0.44	-0.14	0.75	0.70	-0.09	0.24	0.15	-0.09	0.16	0.20	-0.09	0.16	0.15	0.01	0.37	0.43	0.06	0.32	0.32	-0.01
Glycylglycylglycylglycylglycine	0.12	0.18	0.06	0.08	0.07	-0.01	0.66	0.47	-0.19	0.71	0.75	0.04	0.22	0.17	0.01	0.03	0.03	0.01	0.35	0.23	-0.12	0.30	0.37	-0.03

For paper chromatograms developed in one direction only this is about 20, and hence in a two-dimensional chromatogram 20^2 or 400 spots can be separated if their R_F values are suitably different in the two solvent systems used. The best system from the point of view of resolution that we know today is two-dimensional gel electrophoresis on acrylamide gel. Using a pH gradient in one direction and a size exclusion (porosity) gradient in the other, more than two thousand spots of proteins have been separated. See for example Figure 13.

References

[1] A. I. Vogel, *Quantitative Inorganic Analysis*, Longmans, Green & Co., London (1947), p. 182.
[2] H. Irving and R. J. P. Williams, in *Metodi di Separazione nella Chimica Inorganica*, Vol. II, CNR, Rome (1963), pp. 13–62.
[3] *Gmelin Handbook of Inorganic Chemistry, Uranium Supplement volume D2*, Springer Verlag, Berlin (1982), p. 138.
[4] A. J. P. Martin, in *75 Years of Chromatography*, edited by L. S. Ettre and A. Zlatkis, Elsevier, Amsterdam (1979), pp. 285–296.
[5] G. C. Casinovi, in *Metodi di Separazione nella Chimica Inorganica*, Vol. I, CNR, Rome (1963), pp. 7–37.
[6] F. Cramer, *Papierchromatographie*, 2nd edition, Verlag Chemie, Weinheim, (1953), p. 18.
[7] J. Boldingh, *Experientia* **4** (1948) 270; *Recl. Trav. Chim.* **69** (1950) 247.
[8] A. L. LeRosen, *J. Am. Chem. Soc.* **64** (1942) 1905.
[9] A. J. P. Martin, *Biochem. Soc. Symposia* No. 3 (1949) 4.
[10] E. C. Bate-Smith and R.G. Westall, *Biochim. Biophys. Acta* **4** (1950) 427.
[11] I. M. Hais and K. Macek, *Paper Chromatography*, Academic Press, New York and London (1963), p. 69.
[12] T. B. Moore and C. G. Baker, *J. Chromatogr.* **1** (1958) 518.
[13] P. G. Righetti, *Isoelectric Focusing*, Elsevier, Amsterdam (1983), p. 248.

3 PAPER CHROMATOGRAPHY AND THIN LAYER CHROMATOGRAPHY

Paper chromatography of metal ions

I have not yet mentioned the techniques of paper chromatography because, compared with other chromatographic methods, it is so simple that a few words suffice. Usually I could teach new co-workers how to do paper chromatography in half an hour.

Figure 1 shows some types of jar used. One can find something suitable in any laboratory. The solutions to be analysed are placed on the paper by the aid of glass capillaries. Some care must be taken to avoid too much on the paper. The development 'runs itself', taking from 30 minutes to some hours or overnight depending on the solvent system used.

When working with two-phase systems, e.g. butanol saturated with 2 N HCl, a dish of the 'aqueous phase' must be placed in the jar. The atmosphere becomes saturated with all the constituents at their respective vapour pressures. On development, the paper retains the aqueous phase, and the front of the moving liquid would be impoverished with respect to the aqueous phase unless it can regain water from the surrounding atmosphere. With fast or one-phase solvents this is not necessary.

When a solvent system such as butanol–water–hydrochloric acid moves along the paper there is also a frontal analysis effect. The first few centimetres will be only butanol, then some centimetres will be butanol–water and only then will HCl also move along the paper. The exact position of these fronts can be seen when drying the chromatogram, as the first zone dries much faster, then the HCl seems to wash a brown impurity out of the paper which concentrates at the front. Also, spraying with an indicator will show how high the HCl has moved. When mixtures less than about 1 N in

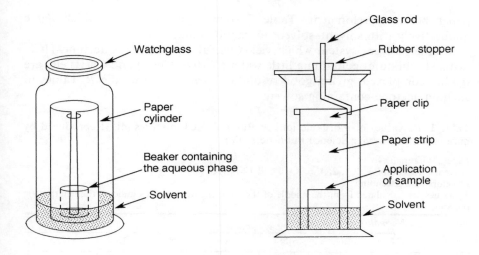

Figure 1. Apparatus for paper chromatographic development

HCl are used as solvent, the HCl front hardly moves at all. Thus the substances chromatograph essentially in butanol–water.

Research on spot tests was generally aimed at finding a specific reagent for a metal ion. In paper chromatography general reagents are needed which detect many metal ions. The most useful are H_2S, ammonium sulphide and 8-hydroxyquinoline. The dried chromatogram is held in H_2S vapour, sprayed with a reagent (with an all-glass nasal spray), or dipped into a solution of the reagent (when the reaction produces an insoluble precipitate).

Metal ions were first separated in 1948 and then almost simultaneously in three different laboratories: by myself at the Sydney Technical College [1], by Arden et al. [2] in London and by Pollard et al. in Bristol [3]. The main work was done in the following 10 years or so. Looking at the solvent extraction data of metal ions into ether from 6 N HCl (see page 10), it seems that this system should furnish good separations especially for As(III), Sb(III) and Sn(II). However, ether is too volatile for paper chromatography, and it does not equilibrate quickly with an aqueous phase on paper strips.

Mixtures of butanol–water–acid proved to give much better chromatograms. Tables 1–6 show typical systems containing HCl, HBr, $HClO_4$ and HNO_3. The metal ions which have high R_F values are usually complexes, viz. $HAuCl_4$, $HBiCl_4$ and $CdCl_2$, whereas non-complexed ions, i.e. aquo-complexes, have low R_F values.

As in solvent extraction the *salting out* principle can be used. One can impregnate filter paper with an aqueous solution of the salt and air dry the paper before development. Table 7 shows typical results, which agree qualitatively with solvent–solvent extraction data.

Another solvent system which yields useful separations is acetone–HCl–water. In mixtures containing little water (5–10%) Co(II) and Cu(II) etc. are readily complexed, giving good resolution. Table 8 shows the R_F values in such an acetone-containing solvent.

Table 1. R_F values of inorganic ions in Butanol–HCl mixtures [4]. (Reproduced by permission of Elsevier Science Publishers BV)

Paper: Whatman No. 1.
Solvents: n-Butanol with 1, 2, 4, 6, 8, 10 and 12 N HCl (1:1)
Development: ascending
R_F values are calculated from the middle of the spot; the other values indicate the length of the spot

Ions	Normality of HCl						
	1N	2N	4N	6N	8N	10N	12N
Group 1: A—Li, Na, K, Rb, Cs; B—Cu, Ag, Au							
Li^+	0.12±0.04	0.16±0.06	0.29±0.08	0.52±0.08	0.49±0.07	0.45±0.07	0.41±0.06
$Na^+(^{22}Na)$	0.04	0.06	0.25	0.39	0.35	0.25	0.20
K^+	0.03±0.03	0.05±0.03	0.19±0.06	0.38±0.05	0.34±0.05	0.25±0.07	0.18±0.08
Rb^+	0.04±0.03	0.06±0.04	0.22±0.05	0.41±0.04	0.39±0.04	0.30±0.04	0.29±0.03
Cs^+	0.04±0.03	0.06±0.04	0.25±0.04	0.45±0.05	0.44±0.04	0.37±0.04	0.37±0.04
Cu^{2+}	0.07±0.03	0.11±0.04	0.31±0.05	0.52±0.05	0.58±0.05	0.57±0.04	0.57±0.04
Ag^+	0	0	0	0	0	0	0
Au^{3+}	0.66±0.02	0.78±0.03	1.00±0.02	0.99±0.01	0.99±0.01	0.97±0.02	0.96±0.02
Group 2: A—Be, Mg, Ca, Sr, Ba, Ra; B—Zn, Cd, Hg							
Be^{2+}	0.12±0.04	0.18±0.04	0.44±0.04	0.59±0.04	0.59±0.04	0.52±0.04	0.42±0.04
Mg^{2+}	0.03±0.01	0.06±0.02	0.23±0.04	0.43±0.04	0.40±0.04	0.35±0.04	0.29±0.04
Ca^{2+}	0.02±0.01	0.03±0.02	0.18±0.03	0.34±0.04	0.33±0.03	0.23±0.04	0.18±0.03
Sr^{2+}	0	±0.01	0.01±0.02	0.08±0.02	0.30±0.04	0.25±0.04	0.16±0.04	0.10±0.03
Ba^{2+}	0	±0.02	0.01±0.01	0.08±0.02	0.17±0.07	0.09±0.09	0.05±0.05	0.03±0.03
$Ra^{2+}(AcX)$	0	0	0.11	0.14	0.08	0.05	0.02
Zn^{2+}	0.60±0.03	0.71±0.02	0.98±0.02	1.00±0.02	0.89±0.03	0.84±0.05	0.75±0.05
Cd^{2+}	0.52±0.04	0.67±0.02	0.97±0.01	0.99±0.01	0.98±0.02	0.91±0.03	0.80±0.04
Hg^{2+}	0.79±0.04	0.84±0.03	1.00±0.01	0.99±0.01	0.98±0.03 (comet to 0.89)	0.90±0.03	0.83±0.04
Group 3: A—Sc, Y, La, Ac; B—B, Al, Ga, In, Tl							
Sc^{3+}	0.02±0.02	0.04±0.03	0.18±0.04	0.40±0.04	0.35±0.04	0.31±0.04	0.30±0.05
Y^{3+}	0.01±0.02	0.03±0.02	0.17±0.03	0.37±0.03	0.31±0.03	0.24±0.03	0.20±0.03
La^{3+}	0.01±0.02	0.02±0.02	0.15±0.03	0.32±0.02	0.26±0.02	0.20±0.02	0.17±0.03
$Ac^{3+}(^{228}Ac)$	0.01	0.04	0.17	0.33	0.27	0.21	0.17
BO_3H_2	0.57±0.05	0.62±0.05	0.76±0.06	0.76±0.06	0.75±0.05	0.67±0.05	0.60±0.06
Al^{3+}	0.01±0.03	0.04±0.03	0.21±0.05	0.41±0.07	0.38±0.07	0.34±0.06	0.27±0.06
Ga^{3+}	0.21±0.02	0.34±0.02	0.76±0.02	0.99±0.02	1.00±0.02	1.00±0.02	0.91±0.04
In^{3+}	0.24±0.08	0.28±0.08	0.45±0.07	0.57±0.08	0.55±0.07	0.49±0.06	0.45±0.07
Tl^{3+}	0.80±0.06	0.94±0.04	0.99±0.01	0.98±0.02	0.98±0.02	0.96±0.02	0.95±0.02
Lanthanides—La, Ce, Pr, Nd, Pm, Sm, Eu, Gd, Tb, Dy, Ho, Er, Tm, Yb, Lu							
La^{3+}	0.01	0.02	0.15	0.32	0.26	0.20	0.17
Ce^{3+}	0.01	0.02	0.17	0.33	0.27	0.22	0.17
Pr^{3+}	0.01	0.02	0.17	0.29	0.28	0.20	0.15
Nd^{3+}	0.01	0.02	0.17	0.31	0.28	0.20	0.16

Table 1. (continued)

Ions	Normality of HCl						
	1N	2N	4N	6N	8N	10N	12N
Pm^{3+}	0.01	0.02	0.18	0.35	0.27	0.19	0.15
Sm^{3+}	0.01	0.02	0.17	0.31	0.26	0.21	0.16
Eu^{3+}	0.01	0.02	0.15	0.30	0.27	0.23	0.18
Gd^{3+}	0.01	0.03	0.19	0.34	0.30	0.24	0.19
Tb^{3+}	0.01	0.03	0.19	0.36	0.34	0.24	0.20
Dy^{3+}	0.01	0.02	0.20	0.37	0.34	0.25	0.20
Ho^{3+}	0.01	0.02	0.19	0.35	0.34	0.25	0.20
Er^{3+}	0.01	0.02	0.19	0.36	0.35	0.27	0.19
Tm^{3+}	0.01	0.01	0.18	0.37	0.34	0.26	0.18
Yb^{3+}	0.01	0.02	0.20	0.36	0.32	0.23	0.16
Lu^{3+}	0.01	0.02	0.19	0.37	0.31	0.23	0.16

Group 4: A—Ti, Zr, Hf, Th; B—Ge, Sn, Pb

	1N	2N	4N	6N	8N	10N	12N
Ti^{3+}	0.06±0.03	0.10±0.02	0.28±0.04	0.47±0.03	0.45±0.04	0.41±0.04	0.44±0.07
Ti^{4+}	0.06±0.03	0.11±0.03	0.30±0.04	0.46±0.04	0.46±0.04	0.44±0.04	0.49±0.05
ZrO^{2+}	0 ±0.02	0.01±0.03	0.10±0.04	0.27±0.07	0.24±0.05	0.17±0.05	0.12±0.07
Hf^{4+}	0.01±0.02	0.02±0.03	0.12±0.04	0.30±0.06	0.27±0.05	0.21±0.04	0.12±0.05
Th^{4+}	0 ±0.01	0.01±0.01	0.17±0.02	0.32±0.03	0.24±0.03	0.16±0.02	0.13±0.02
Ge^{4+}	0.25±0.05	0.32±0.05	0.60±0.07	0.85±0.08	0.96±0.04	0.97±0.03	0.99±0.02
Sn^{2+}, Sn^{4+}	0.67±0.02	0.79±0.02	1.00±0.01	0.98±0.03	0.92±0.04	0.85±0.03	0.80±0.03
Pb^{2+}	0.11±0.02	0.17±0.02	0.39±0.03	0.55±0.02	0.52±0.03	0.45±0.04	0.39±0.04

Group 5: A—V, Nb, Ta, Pa; B—N, P, As, Sb, Bi

	1N	2N	4N	6N	8N	10N	12N
VO^{2+}	0.11±0.08	0.14±0.09	0.26±0.10	0.48±0.09	0.43±0.09	0.36±0.09	0.31±0.07
Nb^{5+}	0	0	0	0	0	0	0
Ta^{5+}	0	0	0	0	0	0	0
$Pa^{5+}(^{233}Pa)$	0	0.01	0.02	0.02	0.03	0.03	0.02
NH_4^+	0.08±0.04	0.11±0.04	0.28±0.05	0.52±0.05	0.51±0.06	0.42±0.05	0.38±0.06
$H_2PO_4^-$	0.55±0.05	0.66±0.04	0.91±0.04	0.93±0.04	0.90±0.05	0.78±0.04	0.77±0.04
As^{3+}	0.52±0.06	0.57±0.06	0.83±0.05	0.96±0.04	0.92±0.03	0.84±0.03	0.77±0.04
As^{5+}	0.62±0.04	0.71±0.05	0.92±0.03	0.97±0.03	0.92±0.03	0.86±0.03	0.77±0.04
Sb^{3+}	0.65±0.03 (comet to 0.54)	0.76±0.02 (comet to 0.67)	0.99±0.01	0.95±0.02	0.83±0.02	0.77±0.03	0.73±0.04
Bi^{3+}	0.56±0.04	0.61±0.05	0.73±0.07	0.79±0.04	0.68±0.05	0.58±0.05	0.49±0.05

Group 6: A—Cr, Mo, W, U; B—S, Se, Te, Po

	1N	2N	4N	6N	8N	10N	12N
Cr^{3+}	0.05±0.04	0.09±0.04	0.22±0.05	0.44±0.04	0.37±0.04	0.30±0.05	0.22±0.05
Mo^{6+}	0.31±0.10	0.45±0.10	0.79±0.08	0.91±0.05	0.91±0.04	0.84±0.05	0.75±0.06
W^{6+}	0	0	0	0	0	0	0
UO_2^{2+}	0.14±0.04	0.20±0.04	0.37±0.05	0.50±0.06	0.51±0.06	0.56±0.06	0.58±0.06
U^{4+}	0.01±0.02	0.02±0.02	0.16±0.03	0.31±0.04	0.24±0.04	0.19±0.03	0.17±0.03
HSO_4^-	0.31	0.44	0.62	0.80	0.87	0.78	0.70
$HSeO_3^-$	0.59±0.04	0.66±0.04	0.81±0.04	0.82±0.05	0.82±0.03	0.97±0.03	0.96±0.04
$HTeO_3^-$	0.15±0.05	0.32±0.05	0.84±0.03	0.96±0.04	0.95±0.05	0.90±0.03	0.91±0.04
$Po^{4+}(^{210}Po)$	0.63	0.79	0.99	0.96	0.95	0.87	0.86

Group 7: A—Mn, Tc, Re; B—Br, I

	1N	2N	4N	6N	8N	10N	12N
Mn^{2+}	0.5 ±0.03	0.11±0.03	0.22±0.07	0.47±0.07	0.46±0.07	0.41±0.06	0.37±0.07
$TcO_4^-(^{99}Tc)$	0.67	0.77	0.98	0.99	0.90	0.73	0.67
$ReO_4^-(^{186}Re)$	0.64±0.02	0.74±0.03	0.98±0.02	0.99±0.02	0.93±0.05	0.88±0.05	0.82±0.05
Br^-	0.63±0.02	0.78±0.02	0.93±0.04	0.91±0.04	0.85±0.04	0.81±0.04	0.75±0.05
I^-	0.68±0.04	0.83±0.05	0.99±0.01	0.96±0.02	0.89±0.02	0.84±0.02	0.79±0.03

(continued overleaf)

Table 1. (*continued*)

Ions	Normality of HCl						
	1N	2N	4N	6N	8N	10N	12N

Group 8: A—Fe, Ru; B—Co, Rh, Ir

Ion	1N	2N	4N	6N	8N	10N	12N
Fe^{2-}	0.05±0.05	0.07±0.05	0.18±0.08	0.45±0.07 0.98±0.02	0.46±0.09 0.98±0.02	0.97±0.03 (comet to 0.40)	0.97±0.03 (comet to 0.76)
Fe^{3+}	0.09±0.04 0.20±0.03	0.14±0.03 0.27±0.06	0.31±0.04 0.51±0.10	0.49±0.05 0.98±0.02	0.39±0.06 0.98±0.02	0.97±0.03 (comet to 0.59)	0.97±0.03 (comet to 0.79)
Ru^{4+}	0.13±0.03 (comet)	0.18±0.03 (comet)	0.33±0.03	0.49±0.03	0.52±0.03	0.44±0.03	0.35±0.03
Co^{2+}	0.03±0.03 0.09±0.03	0.05±0.04 0.15±0.03	0.22±0.05 0.35±0.04	0.42±0.07 0.52±0.04	0.40±0.07 0.53±0.03	0.46±0.07 0.46±0.03	0.63±0.11 0.42±0.03
Rh^{3+}	(comet to 0)	0.11±0.03 0.05±0.02	0.27±0.03 0.19±0.03	0.38±0.04	0.40±0.02	0.31±0.03	0.27±0.02
Ir^{4+}	0.06±0.02	0.12±0.03	0.39±0.04	0.51±0.04	0.56±0.04	0.57±0.04	0.55±0.04
Ni^{2+}	0.03±0.03	0.05±0.03	0.22±0.05	0.41±0.07	0.37±0.07	0.28±0.06	0.24±0.05
Pd^{2+}	0.47±0.04	0.62±0.04	0.80±0.03	0.79±0.04	0.73±0.04	0.62±0.03	0.54±0.05
Pt^{4+}	0.62±0.04	0.73±0.04	0.92±0.03	0.90±0.05	0.87±0.05	0.75±0.06	0.71±0.08 (comet to 0.51)

Table 2. R_F values of metal ions in Butanol equilibrated with various mineral acids [5]. (Reproduced by permission of Elsevier Science Publishers BV)

Paper: Whatman No. 1 and Ekwip No. 1
Development: ascending
R_R values measured with reference to the aqueous front and the metals moving beyond it are given R_F values above I

Metal ion	Solvents			
	Butanol satd. with 1 N HCl	Butanol satd. with 10% HNO_3	Butanol satd. with 10% HBr	Butanol satd. with 20% HBr
Ag^+	0.0	0.23	0.03	
Pb^{2+}	0.0 (tailing)	0.15	0.41	0.44
Hg^{2+}	1.05	tailing	1.25	1.15
Bi^{3+}	0.65	0.27	0.95	0.95
Cu^{2+}	0.10	0.17	0.15	0.16
Cd^{2+}	0.60	0.19	0.95	0.95
As^{3+}	0.70		0.77	0.68
Sb^{3+}	0.8 (tailing)		tailing	tailing
Sn^{2+}	0.95			
Fe^{3+}	0.12	0.18	0.07	
Co^{2+}	0.07	0.17	0.08	
Ni^{2+}	0.07	0.17	0.07	
Mn^{2+}	0.09	0.16	0.10	
Al^{3+}	0.07	0.11	0.14	
Cr^{3+}	0.07	0.15		

Table 2. (*continued*)

Metal ion	Solvents			
	Butanol satd. with 1 N HCl	Butanol satd. with 10% HNO$_3$	Butanol satd. with 10% HBr	Butanol satd. with 20% HBr
Zn^{2+}	0.76	0.15	0.56	
Ca^{2+}	0.03			
Ba^{2+}	0.0	0.08		
Sr^{2+}	0.0	0.08		
Mg^{2+}	0.11			
Na$^+$	0.07			
K$^+$	0.08			
Cs$^+$	0.08	0.13		
Rb$^+$	0.08	0.13		
UO$_2^{2+}$	0.20	0.4	0.15	
Tl$^+$	0.0	0.16	0.02	
Tl^{3+}	1.11		1.18	
MoO$_4^{2-}$	0.5			
Th^{4+}	0.03	0.10		
Be^{2+}	0.30		0.39	
In^{3+}	0.33		0.37	

Table 3. R_F values of inorganic ions in mixtures of Butanol and HBr [6]. (Reproduced by permission of Elsevier Science Publishers BV)

Paper: Whatman No. 1
Solvents: (I) Butanol 100 ml + hydrobromic acid 10 ml + water 90 ml (two phases): the butanol layer was used as solvent.
    (II) Butanol fraction of solvent I + hydrobromic acid 10 ml (one phase).
    (III) Butanol fraction of solvent I + hydrobromic acid 20 ml (one phase).
    (IV) Butanol fraction of solvent I + hydrobromic acid 40 ml (one phase).
    (V) Butanol fraction of solvent I + hydrobromic acid 60 ml (one phase).
The solvents were used after 24 h standing in the chromatography jars.
Development: ascending.
Temperature: 22° ± 2°.

Ion	Solvent				
	I	II	III	IV	V
K	0.04	0.11			
Rb	0.04	0.11			
Cs	0.04	0.11			
Be	0.33	0.46	0.52	0.61	0.66
Mg	0.08	0.16	0.24	0.37	0.45
Ca	0.04	0.08	0.12	0.19	0.25
Sr	0.01	0.04	0.07	0.13	0.15
Ba	0.01	0.02	0.04	0.06	0.08
Zn	0.74	1.00	1.00	1.00	1.00

(*continued overleaf*)

Table 3. (*continued*)

Ion	Solvent				
	I	II	III	IV	V
Cd	0.80	1.00	1.00	1.00	1.00
Hg(II)	1.00	1.00	1.00	1.00	1.00
Cu(II)	0.10	1.00	1.00	1.00	1.00
Ag	0.00	0.10t[a]	0.1–0.86t	0.55–1.00t	0.55–1.00t
Sn(II)	0.83	1.00	1.00	1.00	1.00
Pb	0.40	0.60	0.60	0.64	0.65
As(III)	0.54	0.78	1.00		
Sb	0.90	1.00	1.00	1.00	1.00
Bi	0.80	1.00	0.92	0.85	0.78
Cr	0.03	0.06	0.09	0.23	0.30
Al	0.06	0.16	0.23	0.36	0.40
Fe(III)	0.11	0.58	1.00	1.00	1.00
Ni	0.07	0.13	0.20	0.32	0.33
Co	0.06	0.14	0.24	0.38	0.55
Rh	0.59	0.94	1.00	1.00	1.00
Pd	0.58	0.93	1.00	1.00	1.00
Pt	0.20t	0.95	1.00	1.00t	1.00t
Au	0.67	0.97	1.00	1.00	1.00
Mn(II)	0.09	0.19	0.25	0.32	0.33
Sc	0.02	0.08	0.15	0.30	0.38
Y	0.02	0.07	0.13	0.24	0.30
La	0.00	0.03	0.08	0.17	0.24
Ti	0.07	0.19	0.26	0.43	0.49
V	0.12	0.20	0.27	0.43	0.47
Zr	0.00	0.01	0.03	0.11	0.14
Th	0.00	0.03	0.09	0.15	0.18
U(VI)	0.11	0.21	0.30	0.43	0.47
U(IV)	0.00	0.02	0.06	0.17	0.25
Ga	0.04	0.10	0.26	0.65	0.62
In	0.78	1.00	1.00	1.00	1.00
Tl(III)	0.93	1.00	1.00	1.00	1.00
Ce(III)	0.02	0.04	0.08	0.15	0.23
$B_4O_7^{2-}$	0.60	0.74	0.77	0.80	0.83
$Cr_2O_7^{2-}$	0.11t	0.17	0.22	0.36	0.40
SeO_3^{2-}	0.63	0.78	1.00	1.00	1.00
TeO_3^{2-}	0.18	0.91	1.00	1.00	1.00
MoO_4^{2-}	0.21	0.38	0.48	0.53	0.55
PO_4^{3-}	0.53	0.82	0.90	0.96	0.96
ReO_4^-	0.64	0.94	1.00	1.00	1.00
WO_4^{2-}	0.00	0.00	0.00	0.00	0.00
$[Fe(CN)_6]^{4-}$	0.64	1.00	1.00	1.00	1.00
$[Fe(CN)_6]^{3-}$	0.52t	0.91			
I^-	0.68	0.99	1.00	1.00	1.00
ClO_3^-	0.62	0.98	1.00	1.00	1.00

[a] t = tailing.

Table 4. R_F values of cations in alcohol–HNO_3 mixtures [7]

Paper: Whatman No. 1.
Development: descending.

Cation	Solvent		
	Butanol satd. with 1 N HNO_3	Butanol satd. with 2 N HNO_3	Butanol satd. with 3 N HNO_3
Ag^+	0.12	0.13	0.19
Pb^{2+}	0.01	0.08	0.15
Bi^{3+}	0.18	0.19	0.25
Cu^{2+}	0.10	0.11	0.15
Cd^{2+}	0.10	0.11	0.15
As^{3+}	0.45	0.45	0.48
Sn^{2+}	0.74	0.76	0.84
Sn^{4+}	0.65	0.68	0.75
Al^{3+}	0.05	0.05	0.12
Cr^{3+}	0.06	0.06	0.15
Fe^{3+}	0.06	0.06	0.15
Zn^{2+}	0.05	0.06	0.15
Mn^{2+}	0.08	0.09	0.16
Co^{2+}	0.08	0.09	0.16
Ni^{2+}	0.05	0.06	0.15
Ca^{2+}	0.06	0.06	0.14
Sr^{2+}	0.05	0.05	0.14
Ba^{2+}	0.05	0.06	0.13
Mg^{2+}	0.07	0.09	0.18
Na^+	0.06	0.07	0.15
K^+	0.06	0.07	

Table 5. R_F values of some transuranic elements and rare earths [8]

Paper: Whatman No. 1
Development: ascending
Temperature: 20°.
All values indicated in the table are the averages of four measurements

Ion	Molarity of HNO_3 in n-butanol–HNO_3 (1:1) mixtures									
	0 M	0.5 M	1 M	2 M	3 M	3.5 M	4 M	4.5 M	5 M	6 M
Pu(III)			0.049	0.081	0.156	0.179	0.250	0.311	0.360	0.342
Pu(IV)			0.053	0.149	0.267	0.327	0.389	0.478	0.528	0.581
Pu(VI)[a]	⎰ 0	⎰ 0.118	⎰ 0.169	⎰ 0.254	⎰ 0.345	⎰ 0.389	⎰ 0.455			
	⎱ 0	⎱ 0.036	⎱ 0.076	⎱ 0.160	⎱ 0.256	⎱ 0.312	⎱ 0.369	0.467	0.534	0.578
Am(III)			0.043	0.072	0.150	0.168	0.233	0.329	0.352	0.341
U(VI)	0.099	0.205	0.256	0.368	0.447	0.496	0.513	0.533	0.593	0.630
Th(IV)		0.017	0.038	0.076	0.141	0.169	0.230	0.270	0.347	0.344
La(III)			0.043	0.077	0.132	0.169	0.208	0.242	0.331	0.344
Ce(III)		0.026	0.043	0.075	0.141	0.177	0.252	0.267	0.357	0.371
Nd(III)			0.050	0.075	0.131	0.164	0.212	0.272	0.340	0.339
Pm(III)			0.043	0.077	0.149	0.196	0.259	0.375	0.370	0.344

(continued overleaf)

Table 5. (*continued*)

Ion	Molarity of HNO₃ in n-butanol–HNO₃ (1:1) mixtures									
	0 M	0.5 M	1 M	2 M	3 M	3.5 M	4 M	4.5 M	5 M	6 M
Sm(III)			0.036	0.069	0.128	0.184	0.237	0.292	0.347	0.333
Eu(III)			0.049	0.076	0.134	0.176	0.219	0.280	0.326	0.351
Gd(III)			0.036	0.067	0.130	0.178	0.244	0.293	0.355	0.357

[a] Pu(VI) gives two close spots in 0.5–4 M.

Table 3.6. R_F values of metal ions in Butanol–aqueous HClO₄ mixtures. (Reprinted from Ref. 9, p. 237 by courtesy of Marcel Dekker Inc.)

Paper: Whatman 3MM
Solvents: Butanol equil. with an equal volume of aqueous HClO₄ of the normality indicated in the table

Metal ion	Normality of aqueous HClO₄				
	0.1	0.5	1	2	2.25
Li	0.27	0.27	0.27	0.39	0.47
Na	0.10	0.10	0.11	0.27	0.33
Ag	0.16t	0.17t	0.23t	0.27t	0.28t
Hg(I)	0.12	0.17	0.23	0.34	0.36
Tl(I)	0.04	0.07	0.10	0.22	0.29
Be	0.31	0.37	0.43	0.47	0.53
Mg	0.10	0.11	0.17	0.26	0.37
Ca	0.12	0.11	0.18	0.27	0.30
Sr	0.07	0.07	0.11	0.19	0.28
Ba	0.03t	0.04	0.07	0.16	0.23
Al	0.08	0.09	0.15	0.27	0.39
Zn	0.10	0.11	0.17	0.27	0.37
Ga	0.06	0.13	0.22	0.37	0.47
In	0.02	0.11	0.20	0.38	0.44
Mn	0.11	0.18	0.24	0.37	0.42
Fe(III)	0.09t	0.12	0.17	0.34	0.43
Ni	0.12	0.15	0.16	0.31	0.36
Co	0.11	0.13	0.15	0.32	0.40
Cu	0.12	0.14	0.17	0.33	0.39
Cd	0.14	0.16	0.17	0.32	0.36
Hg(II)	0.09t	0.13t	0.19t	0.27t	0.32t
Pb	0.05t	0.05t	0.07t	0.14t	0.18t
Bi	0.00	0.02t	0.11	0.19	0.25t
UO_2^{2+}	0.17	0.26	0.26	0.35	0.40
Y	0.01t	0.06	0.06	0.23	0.24
La	0.08t	0.09	0.15	0.30	0.32
Pr	0.01t	0.04	0.07	0.18	0.26
Nd	0.02t	0.05	0.13	0.25	0.27
Sm	0.03t	0.05	0.13	0.26	0.29
Eu	0.01t	0.05	0.06	0.20	0.24
Gd	0.02t	0.06	0.09	0.24	0.25
Tb	0.02t	0.07	0.06	0.23	0.24

Table 6. (*continued*)

Metal ion	Normality of aqueous $HClO_4$				
	0.1	0.5	1	2	2.25
Yb	0.02t	0.07	0.04	0.23	0.23
Sc	0.06	0.11	0.16	0.30	0.39
Ti	0.00	0.00	0.00	0.39	0.48
	0.06	0.16t	0.26		
Zr	0.00	0.02t	0.07t	0.16	0.23
Th	0.06t	0.08	0.15	0.29	0.37
SeO_3^{2-}	0.56	0.59	0.59	0.63	0.64
TeO_3^{2-}	0.12	0.16	0.18	0.32	0.33
TcO_4^-	0.64	0.79	0.87	0.92	0.93
ReO_4^-	0.64	0.79	0.87	0.92	0.93
MoO_4^{2-}	0.16	0.41	0.51	0.60	0.62

t = tailing.

Table 7. R_F values of certain cations with Butanol–1.5 N HNO_3 on papers impregnated with salts [10]. (Reproduced by permission of Elsevier Science Publishers BV)

Paper: d'Arches paper No. 302
Development: ascending at room temperature

Cation	No impregnation	Paper impregnated with[a]		
		KNO_3	NH_4NO_3	$NaNO_3$
UO_2^{2+}	0.30	0.50	0.74	0.88–0.99
Fe^{3+}	0.05	0.06	0.12	0.12
Cu^{2+}	0.05	0.08	0.07	0.10
Co^{2+}	0.07	0.07	0.07	0.10

[a] 1 N, aqueous; subsequent drying.

Table 8. R_F values of inorganic ions in acetone–HCl–water mixtures [11]. (Reproduced by permission of Elsevier Science Publishers BV)

Paper: Whatman No. 1
Solvent: Acetone–water–HCl, $d = 1.124 (40:10:5)$
Development: ascending

Ion	R_F	Ion	R_F	Ion	R_F
Ag^+	0.920	As^{3+}	0.785	Al^{3+}	0.211
Hg^{2+}	0.992	Bi^{3+}	0.970	Cr^{3+}	0.255
Pb^{2+}	0.732	Sb^{3+}	0.984	Ti^{4+}	0.291
Zn^{2+}	0.977	Sn^{2+}	0.984	Th^{4+}	0.082
Mn^{2+}	0.347	Pd^{2+}	0.972	MoO_4^{2-}	0.940
Ni^{2+}	0.167	Pd^{2+}	0.568	VO_3^{2-}	0.360
Co^{2+}	0.260	ZrO^{2+}	0.017	CrO_4^{2-}	0.970
Cu^{2+}	0.655	UO_2^{2+}	0.944	WO_4^{2-}	0.985
Cd^{2+}	0.943	Fe^{3+}	0.984	Tl^+	0.008

Data for many solvent systems are available in the literature and good collections have been published [12].

We shall now discuss some specific and interesting separations.

THE SEPARATION OF LITHIUM, SODIUM AND POTASSIUM [2]

A neutral solution of the three chlorides separates with pure methanol as solvent (R_F values KCl 0.22, NaCl 0.44, LiCl 0.72). The amount of chloride associated with each cation is stoichiometric. This permits a simple detection by dipping into silver nitrate solution and exposure to UV light. The three cations can also be determined by an argentometric titration. Various authors have tried to improve on this separation with ethanol–water mixtures in place of methanol, but without success. The separation seems to involve an ion pair association of the cations and the chloride ion, which is remarkable, as alkalis do not generally form ion pairs.

The usual non-chromatographic qualitative tests for the alkalis are not very satisfactory, and thus this separation represents an interesting enrichment of our analytical methods.

If instead of the chlorides neutral chromates are chromatographed, one obtains three well separated yellow spots. However, excess of chromate or acidity or alkalinity mars this separation.

THE SEPARATION OF CAESIUM, RUBIDIUM AND POTASSIUM

A splendid separation of these three ions was obtained using phenol equilibrated with 2 N HCl as solvent [13]; R_F values were K 0.19, Rb 0.27 and Cs 0.43. To my knowledge this is the only system in which the difference for Cs – Rb is greater than that of Rb – K and it gives a practically specific separation of caesium. Why the separability is inverted is not clear. Ion pair formation between phenate and alkali metals seems unlikely in the presence of 2 N HCl. Attempts to improve this separation by changing the solvent were unsuccessful [14].

Some applications

In analytical chemistry we have numerous alternatives to paper chromatography, such as spectrography, polarography and spot tests, so that one has to ask whether there are any uses for paper chromatography. We shall summarize some fields where it has proved useful.

GEOCHEMICAL PROSPECTING

Several geochemical laboratories, faced with large sample numbers, found that paper chromatography had several features which made it the preferred method.

It is very cheap and needs no special skills. It is also not over-sensitive nor too insensitive. One can thus work out specific tests for uranium or nickel or gold which if positive also show that the sample has a high enough content to warrant further investigation. A book on this topic was published by A.S.Ritchie [15] and numerous specific methods were worked out by Agrinier.

PREPARATION OF CARRIER-FREE TRACERS IN RADIOCHEMISTRY

For the separation of radioelements in general and carrier-free tracers in particular, paper chromatography has the advantage that no apparatus is contaminated. The radioactive substances can be dissolved in a microbeaker and placed on the paper with a glass capillary. The desired zone can be eluted or ashed and again only a small beaker will be used.

In this field one prerequisite is that the R_F value of the metal ion does not change at high dilutions. This is the case when solvents containing strong acids (e.g. those in Tables 1–6) are used which suppress the ionization of the few carboxyl groups which cellulose inevitably contains. Solvents such as butanol–acetic acid–water are not suitable.

A lot of work was done in the period 1950–1960 and two small monographs were published [16, 17].

THE SYNTHESIS OF COORDINATION COMPOUNDS

Paper chromatography has been applied in numerous problems in coordination chemistry and we shall illustrate the possibilities with some examples of our work.

The reaction

$$[Pt(NH_3)_4]Cl_2 \rightarrow trans\,[Pt(NH_3)_2Cl_2]^0 + 2NH_3$$

can be achieved by heating solid $[Pt(NH_3)_4]Cl_2$; As intermediate, $[Pt(NH_3)_3Cl]^+ Cl^-$ is formed.

We wanted to establish whether there is an optimum temperature for the formation of the intermediate. We used a Kofler heating bank on which we established a temperature gradient from 140–230°C. The solid $[Pt(NH_3)_4]Cl_2$ was spread on the surface as a thin layer and heated for an hour. Sections corresponding to a 10°C temperature interval were removed, dissolved in water and chromatographed.

The results are shown in Figure 2. There seems to be no temperature region where the intermediate $[Pt(NH_3)_3Cl]Cl$ accumulated to any marked extent.

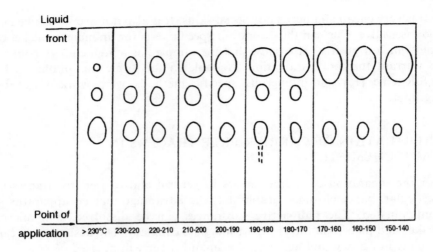

Figure 2. Chromatograms (solvent: 2 N HCl, paper: Whatman 3MM) of [Pt(NH$_3$)$_4$]Cl$_2$ heated on the Kofler heating bank for one hour. Samples corresponding to a 10°C range were analysed. The spot closest to the liquid front is unchanged [Pt(NH$_3$)$_4$]Cl$_2$, the second spot is [Pt(NH$_3$)$_3$Cl]$^+$ and the slowest is trans-[Pt(NH$_3$)$_2$Cl$_2$]^0 [18]

Mixed dihalogeno complexes of the type [Pt(NH$_3$)$_2$XY]^0 can be prepared by the reaction of the dichloro complex with one equivalent of AgNO$_3$, after which the precipitated AgCl is filtered off and one equivalent of either KBr or KI is added.

Analytical data of the isolated compounds can be misleading, as a mixture of equimolecular amounts of [Pt(NH$_3$)$_2$Cl$_2$]^0 and [Pt(NH$_3$)$_2$Br$_2$]^0 will give the same analysis as [Pt(NH$_3$)$_2$ClBr]^0.

Figure 3 shows the separations obtained with acetone–water as solvent. The chlorobromo compounds (*cis* as well as *trans*) were usually pure within the limits of detectability. On the other hand, the presumed chloroiodo complexes usually yielded three spots of about equal intensity.

PREPARATIVE SEPARATIONS

Using sheets of thick Whatman 3MM paper, Dwyer et al. [20] performed preparative separations of Co(III) complexes. The sample is applied as a thin line all along a sheet 20–30 cm wide, and quantities of the order of 20 mg can be separated with the same efficiency as in analytical chromatograms. Dwyer et al. separated mixed [Co^{III}(en)$_x$(pn)$_y$]$^{3+}$ into distinct bands, which were then eluted and recovered. Butanol–HCl–water and butanol–HClO$_4$–water mixtures were used. The amounts that can be handled vary

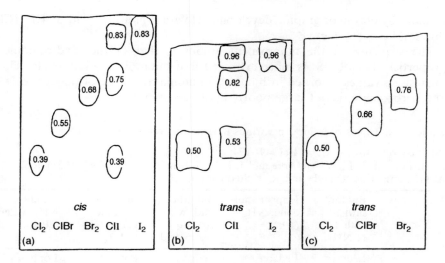

Figure 3. Some typical chromatograms obtained with acetone–water (9:1) at 20 °C on Whatman 3MM paper [19]. Bottom line: line of application; top line: liquid front. (a) Samples of cis-[Pt(NH$_3$)$_2$X$_2$]^0 run side by side on the same sheet. Note that cis-ClI is impure, containing also cis-Cl$_2$ and I$_2$ complexes. (b) Samples of trans-[Pt(NH$_3$)$_2$X$_2$]^0 run side by side on the same sheet. Note that trans-ClI contains also trans-Cl$_2$ and I$_2$ complexes. (c) Samples of trans-[Pt(NH$_3$)$_2$X$_2$]^0 run side by side on the same sheet. (Reproduced by permission of Elsevier Science Publishers BV)

from solvent to solvent as well as from paper to paper. We obtained good separations with acetone–HCl–water mixtures even when 100 mg amounts were chromatographed.

Paper chromatography of inorganic anions

The separation of the usual inorganic anions was first tried with butanol–ammonia mixtures [21]; See Table 9. Later isopropanol–ammonia mixtures were also investigated, and did not yield very different separations. The remarkable feature of these solvents is that polyvalent anions stay at $R_F = 0$ and only monovalent ions migrate.

This is actually due to the calcium content of commercial filter papers. The cellulose is washed during the manufacturing process with large amounts of (hard) water and thus all the carboxylic groups of the paper are converted to the calcium form. Polyvalent anions thus produce insoluble calcium salts at the point of application of the chromatogram. The phenomenon was first reported by Hanes and Isherwood [22], who observed it in separations of orthophosphate and sugar phosphates. By washing the paper (usually by chromatographic development) with 2 N acetic acid or 1 NHCl this can be overcome.

The sequence of the monovalent anions is that of the 'hydrophobic' properties. Thus CNS$^-$ is more extracted into butanol than iodide etc. The highest R_F values recorded in butanol–ammonia are those of pertechnetate and perbromate, i.e. of large poorly hydrated anions.

Table 9. R_F values of anions in various partition solvents
(M. Lederer, *Aust. J. Sci.* **11** (1949) 174;
F. H. Pollard, J. F. W. McOmie and I. I. M. Elbeih, *J. Chem. Soc.* (1951) 466;
A. G. Long, J. R. Quayle and R. J. Stedman, *J. Chem. Soc.*, (1951) 2197)

Metal ions	Butanol satd. with 1.5 N NH$_3$	Isopropanol–conc. NH$_3$ (10:1)	Butanol satd. with 20% aq. acetic acid	Butanol–pyridine–1.5 N NH$_3$ (2:1:2)	Ethanol–NH$_3$–water (80:4:6)
	Lederer	Lederer	Lederer	Pollard et al.	Long et al.
CNS$^-$	0.45	0.8	0.44	0.56	
I$^-$	0.30	0.6	0.32	0.47	0.53
NO$_3^-$	0.24		0.25	0.40	0.48
AsO$_2^-$	0.21		0.41	0.19	
NO$_2^-$	0.20			0.25	0.47
Br$^-$	0.16	0.46	0.16	0.36	0.48
BrO$_3^-$	0.13			0.25	
Cl$^-$	0.10	0.37	0.21	0.24	0.43
IO$_3^-$	0.03			0.09	
F$^-$	0.0				0.27
Co(NO$_2$)$_6^{3-}$	0.0				
S^{2-}	0.0				
S$_2$O$_3^{2-}$	0.0				
IO$_4^-$	0.0				
CrO$_4^{2-}$	0.0	0.01	0.24	0.0	0.09
PO$_4^{3-}$	0.0			0.04	0.04
AsO$_4^{3-}$	0.0		0.15	0.05	
Fe(CN)$_6^{3-}$	0.0	0.01	0.03		
Fe(CN)$_6^{4-}$	0.0	0.01			
ClO$_3^-$				0.42	
CO$_3^{2-}$				0.06	
SO$_4^{2-}$				0.07	0.09
SO$_3^{2-}$					0.04

CONDENSED PHOSPHATES

The elucidation of the solution chemistry of phosphates in solution is one of the great achievements of paper chromatography in inorganic chemistry. J.P.Ebel, who had done the pioneering work in this field, also wrote an excellent review [23]. The details are given in the chapter on phosphates on pages 163–170.

POLYTHIONATES

These ions can be separated by paper chromatography, as shown in Table 10. Excellent separations were also obtained by electrophoresis. See page 171.

Table 10. R_F values of higher polythionic acids in Wackenroder liquid, [27]

Paper: Whatman No. 1
Solvent: n-Butanol–acetic acid–acetoacetic ester – water (10:2:1:7)

Acid	R_F value
$H_2S_4O_6$	0.15 diffuse
$H_2S_5O_6$	0.20
$H_2S_6O_6$	0.25
$H_2S_7O_6$	0.30
$H_2S_8O_6$	0.37
$H_2S_9O_6$	0.45
$H_2S_{10}O_6$	0.52
$H_2S_xO_6$ ($x > 10$)	0.65 diffuse
H_2S	0.90

Paper adsorption chromatography

Metal ions are poorly adsorbed on cellulose with the exception of the very 'hydrophobic' ones, which are also readily extracted into ether. R_F values decrease with increasing HCl or LiCl concentration in the eluent [28]; see Figure 4.

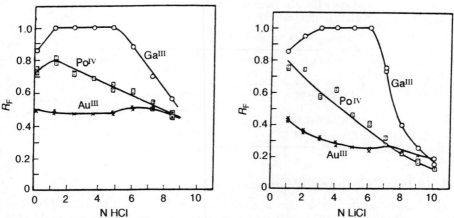

Figure 4. (a) Variation of the R_F values of Ga(III), Po(IV) and Au(III) with the concentration of HCl used as eluent (b) Variation of the R_F values of Ga(III), Po(IV) and Au(III) with the concentration of LiCl used as eluent

Neutral complexes are often adsorbed on paper and good separations have been obtained with Pt(II) ammines; see Tables 11 and 12.

Organic complexes such as thiourea complexes are often adsorbed on paper, and this can form the basis of simple separations. For example, a fast separation of Tc and Re is possible by adding thiourea to a solution in dilute nitric acid containing pertechnetate and perrhenate. The Tc forms a yellow Tc(III) thiourea complex which moves with an R_F of 0.2, while Re is not reduced or complexed and moves faster as ReO_4^- (R_F 0.7) [29].

Sulphuric acid and phosphoric acid produce a very strong salting-out effect with hydrophobic halocomplexes. In 15% H_2SO_4 a splendid separation of Au(III)–Tl(III)–Hg(II) as their bromo-complexes can be obtained in about 20 minutes [30]; see Table 13.

Table 11. R_F values of chloroammineplatinum (II) complexes [18]

Development: ascending

Complex	Solvent and paper	
	0.1 N HCl; Whatman No. 3MM	2N HCl; Whatman No. 3MM
$Pt(NH_3)_4^{2+}$	0.92	0.79
$Pt(NH_3)_3Cl^+$	0.69	0.66
cis-$Pt(NH_3)_2Cl_2^0$	0.70	0.67
$trans$-$Pt(NH_3)_2Cl_2^0$	0.45–0.50	0.48
$PtCl_4^{2-}$	0.91	–
$PtCl_6^{2-}$	0.91	–

Table 12. R_F values of chloropyridineplatinum (II) complexes [18]

Development: ascending

Complex	2N HCl; Whatman No. 3MM paper
$[PtPy_4]Cl_2$	0.96
cis-$PtPy_2Cl_2$	0.87
cis-$Pt(NH_3)_2Cl_2$	0.67
$trans$-$Pt(NH_3)_2Cl_2$	0.48

Thin layer chromatography

The first good thin layer chromatograms were published by Kirchner *et al.* [31] in 1951. The technique became popular around 1956 when Stahl [32], Demole [33] and others introduced standardized adsorbents and published many excellent chromatograms.

Table 13. R_F values of the bromo complexes of Hg(II), Tl(III) and Au(III) in Whatman 3MM with different concentrations of H_2SO_4 as eluent [30]. (Reproduced by permission of Elsevier Science Publishers BV)

Species	H_2SO_4 concentration (% v/v)			
	2.5	5	10	20
Hg(II)	0.72	0.69	0.60	0.41
Tl(III)	0.52	0.50	0.39	0.24
Au(III)	0.22	0.18	0.15	0.06

Today many authors claim that TLC was introduced because it yielded better separations with higher plate numbers in a shorter time than was possible in paper chromatography. The real reason was rather that adsorption chromatograms on adsorbent-impregnated papers often proved unsatisfactory. Also, many lipophilic compound groups are resolved readily by adsorption chromatography on silica or alumina but not so easily in the usual liquid–liquid systems.

TLC of inorganic compounds started around 1960 (for a review see Ref. 34). Some authors achieved separations with very short development times and this is of interest when labile complexes have to be separated. Extensive work by Merkus [35] showed that most separations can be done faster, but not many better, than on paper chromatograms.

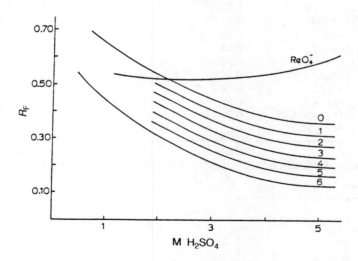

Figure 5. Dependence of R_F values of $ReBr_nCl_{6-n}^{2-}$ complexes and ReO_4^- on the molarity of H_2SO_4 [38]. Numbers indicate value of n. Run of 16 cm in 2.5 h.
Note: With 3.2 M H_2SO_4, both the $ReBr_nCl_{6-n}^{2-}$ and the $OsBr_nCl_{6-n}^{2-}$ series can be completely separated; run, 25–30 cm in 5 h. For quantitative evaluation of the results, a 24 h run (40 cm + overelution) is recommended

An important feature of TLC is that one can chromatograph on many adsorbents, inorganic ion exchangers etc; i.e. there is a much wider range of possibilities than on paper.

A good collection of TLC data for inorganic substances was published by Brinkman et al. [36]. Some R_F values of complexes are shown in Tables 14(a)–(c).

Table 14(a). [37]Comparison of R_F values of some cobalt(III) amine complexes

Stationary phase: Silica gel (pre-coated plates, Type Q-I; Quantum Ind.)
Mobile phase: Formamide–methanol–glacial acetic acid (40:60:0.1)
Conditions: Run, 12 cm in 100 min

Complex	R_F	
	trans	cis
$[Co(en)_2(OAc)_2]^+$	0.63	–
$[Co(en)_2(N_3)_2]^+$	0.85	0.77
$[Co(en)_2(N_3)(NCS)]^+$	0.92	0.85
$[Co(en)_2(N_3)(NO_2)]^+$	0.81	–
$[Co(en)_2Br_2]^+$	0.84	0.77
$[Co(en)_2Cl_2]^+$	0.80	0.77
$[Co(en)_2Cl(NCS_2)]^+$	0.94	0.85
$[Co(en)_2Cl(NO_2)]^+$	0.80	0.75
$[Co(en)_2F_2)]^+$	0.82	0.78
$[Co(en)_2(NCS)_2]^+$	0.95	–
$[Co(en)_2(NCS)(NO_2)]^+$	–	0.78
$[Co(en)_2(H_2O)(OAc)]^{2+}$	0.57	–
$[Co(en)_2(H_2O)Br]^{2+}$	0.58	0.36
$[Co(en)_2(H_2O)Cl]^{2+}$	0.64	0.55
$\{[Co(en)_2(H_2O)F]^{2+}$	0.63	0.41
$[Co(en)_2(H_2O)(NO_2)]^{2+}$	0.70	0.43
$[Co(en)_2(H_2O)(NO_2)]^{2+}$	0.57	0.34
$[Co(en)_2(H_2O)_2]^{3+}$	0.19	0.12
$[Co(NH_3)_4(H_2O)Cl]^{2+}$	0.63	–
$[Co(NH_3)_4(H_2O)(NO_2)]^{2+}$	0.72	0.65
$[Co(NH_3)_4(H_2O)_2]^{3+}$	0.25	0.11

Table 14(b). R_F values of substituted pentamminecobalt(III) complexes

Conditions as in Table 14(a)

Complex	R_F
$[Co(NH_3)_5N_3]^{2+}$	0.60
$[Co(NH_3)_5Br]^{2+}$	0.68
$[Co(NH_3)_5Cl]^{2+}$	0.67
$[Co(NH_3)_5(CN)]^{2+}$	0.55
$[Co(NH_3)_5F]^{2+}$	0.59
$[Co(NH_3)_5(NO_2)]^{2+}$	0.62
$[Co(NH_3)_6]^{3+}$	0.62
Co^{2+}	1.00 T

Table 14(c). Selected R_F values of [Co(en)$_2$AB]$^+$ and their hydrolysis products
Conditions as in Table 14(a)

Complex	R_F	Complex	R_F
trans-Cl$_2$	0.80	trans-NO$_2$–Cl	0.80
cis-Cl$_2$	0.77	cis-NO$_2$–Cl	0.75
trans-Cl–H$_2$O	0.64	trans-NO$_2$–H$_2$O	0.57
cis-Cl–H$_2$O	0.55	cis-NO$_2$–H$_2$O	0.34
trans-(H$_2$O)$_2$	0.19	trans-(H$_2$O)$_2$	0.19
cis-(H$_2$O)$_2$	0.11	cis-(H$_2$O)$_2$	0.11
trans-Br$_2$	0.84	trans-NCS–Cl	0.94
cis-Br$_2$	0.77	cis-NCS–Cl	0.85
trans-Br–H$_2$O	0.58	trans-NCS–H$_2$O	0.70
cis-Br–H$_2$O	0.36	cis-NCS–H$_2$O	0.43
trans-(H$_2$O)$_2$	0.19	trans-(H$_2$O)$_2$	0.19
cis-(H$_2$O)$_2$	0.11	cis-(H$_2$O)$_2$	0.11

Splendid separations of mixed halide complexes of Os(IV) and of Re(IV) were obtained on 40 cm long cellulose thin layers using 3.2 M aqueous H$_2$SO$_4$ as developing solvent [38]; see Figure 5. These separations cannot be achieved on paper chromatograms.

The series of chloro-aquo-Ir(III) species was separated by TLC as shown in Table 15 and Figure 6 [39].

Table 15. R_F Data for Ir(IV) aquohalo complexes

Stationary phases: S_1 = Silica gel MN N-HR.
S_2 = Cellulose MN300 and MN300 HR.
S_3 = Cellulose (pre-coated plates, Merck).
Mobile phases: M_1 = 1-Butanol–methanol–water–acetic acid (75–60:10:10–25:5).
M_2 = Ethanol–water–trichloroacetic acid–Cl$_2$ (60:40:1g:trace).
M_3 = Ethyl acetate–ethanol–water–trichloroacetic acid (85:15:2g).
M_4 = Dioxane–2 N HCl–trichloroacetic acid–Cl$_2$ (90:10:0.5g:trace).
Conditions: Sandwich chamber.
Run in 30–120 min.

Ir(IV) complex	R_F		
	S_2/M_2	S_1/M_3	S_3/M_4
IrCl$_6^{2-}$	0.58	0.36	0.28
IrCl$_5$(H$_2$O)$^-$	0.42	0.53	0.44
IrCl$_4$(H$_2$O)$_2$	0.32	0.78	0.70
IrCl$_3$(H$_2$O)$_3^+$	0.22	–	0.99

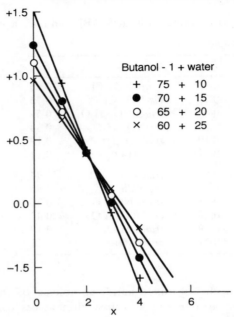

Figure 6. Relation between R_M values and x for TLC of $[\text{Ir(III)Cl}_{6-n}(\text{H}_2\text{O})_x]^{-3+x}$ in the system S_1/M_1 (see Table 15) [39].
Note: A two-dimensional combination of TLC and thin layer electrophoresis has been used to study the behaviour of a solution of irradiated $K_3\text{IrCl}_6$

References

[1] M. Lederer, *Nature* **162**(1948) 776.
[2] T. V. Arden, F. H. Burstall, G. R. Davies, J. A. Lewis and R. P. Linstead, *Nature* **162**(1948) 691.
[3] F. H. Pollard, J. F. W. McOmie and I. I. M. Elbeih, *Nature* **163**(1949) 292.
[4] R. A. Guedes de Carvalho, *Anal. Chim. Acta* **16**(1957) 555.
[5] M. Lederer, *Anal. Chim. Acta* **4**(1950) 629.
[6] S. Kertes and M. Lederer, *Anal. Chim. Acta* **15**(1956) 543.
[7] I. I. M. Elbeih, J. F. W. McOmie and F. H. Pollard, *Discuss. Faraday Soc.* **7**(1949) 183.
[8] F. Clanet, *J. Chromatogr.* **7**(1962) 373.
[9] M. Mazzei and M. Lederer, *Anal. Lett.* **1**(1968) 237.
[10] M. Lederer, *Anal. Chim. Acta* **7**(1952) 458.
[11] A. K. Majumdar and M. M. Chakrabartty, *Anal. Chim. Acta* **17**(1957) 415.
[12] M. Lederer and C. Majani, *Chromatogr. Rev.* **12**(1970) 239.
[13] A. E. Steel, *Nature* **173**(1954) 315.
[14] R. J. Magee and J. H. Headridge, *Analyst* **82**(1957) 95.
[15] A. S. Ritchie, *Chromatography in Geology*, Elsevier, Amsterdam (1964), 185 pp.
[16] M. Lederer, *Chromatographie sur Papier des Radioélements*, Hermann & Cie, Paris (1956), 64 pp.
[17] R. A. Bailey, *Paper Chromatography and Electromigration Techniques in Radiochemistry*, Nuclear Science Series, Radiochemical Techniques, Department of Commerce, Washington, DC (1962), 48 pp.

[18] F. Basolo, M. Lederer, L. Ossicini and K. H. Stephen, *Ric. Sci.*, Ser. 2 **32**(1962) 485.
[19] F. Basolo, M. Lederer, L. Ossicini and K. H. Stephen, *J. Chromatogr.* **10**(1963) 262.
[20] F. P. Dwyer, T. E. McDermott and A. M. Sargeson *J. Am. Chem. Soc.* **85**(1963) 2913.
[21] M. Lederer, *Science* **110**(1949) 115.
[22] C. S. Hanes and F. A. Isherwood, *Nature* **164**(1949) 1107.
[23] J. P. Ebel, in *Metodi di Separazione nella Chimica Inorganica*, Vol. I, CNR, Rome (1963), pp. 199–248.
[24] J. P. Ebel, *Mikrochim. Acta* (1954) 679.
[25] E. Thilo and H. Grunze, *Die Papierchromatographie der kondensierten Phosphate*, Akademie-Verlag, Berlin (1955).
[26] K. Gassner, *Mikrochim. Acta* (1957) 594.
[27] R. Barbieri and M. Bruno, *J. Inorg. Nucl. Chem.* **14**(1960) 148.
[28] M. Lederer, in *Metodi di Separazione nella Chimica Inorganica*, Vol. I, CNR, Rome (1963), pp. 109–116.
[29] T. J. Beckmann and M. Lederer, *J. Chromatogr.* **5**(1961) 341.
[30] M. Lederer, *Anal. Chim. Acta* **246**(1991) 451.
[31] J. G. Kirchner, J. M. Miller and G. J. Keller, *Anal. Chem.* **23**(1951) 420.
[32] E. Stahl, G. Schröter, G. Kraft and R. Renz, *Pharmazie* **11**(1956) 633.
[33] E. Demole, *Compt. Rend.* **243**(1956) 1883.
[34] M. Lederer, *Chromatogr. Rev.* **9**(1967) 115.
[35] F. W. H. M. Merkus, Doctoral Thesis, University of Amsterdam, Alberts Drukkerijen, Sittard (1966), 122 pp.
[36] U. A.Th. Brinkman, G. De Vries and, R. Kuroda, *J. Chromatogr.* **85**(1973) 187.
[37] R. B. Hagel and L. F. Druding, *Sep. Sci.* **4**(1969) 89.
[38] H. Müller, *Fresenius' Z. Anal. Chem.* **247**(1969) 145.
[39] W. J. van Ooij and J. P. W. Houtman, *Fresenius' Z. Anal. Chem.* **236**(1968) 407.

4 ELECTROPHORESIS

Paper Electrophoresis

Precise analytical work using electromigration starts with the work of Tiselius around 1925. A U-tube apparatus was perfected to permit the measurement of moving boundaries. This resulted in the discovery of α, β and γ globulins in human serum. However, no complete separation could be effected as the apparatus yielded only an equivalent of 'frontal analysis'.

The use of paper strips as anticonvection medium for electromigration was, it seems, first reported by Von Klobusitzky and König in 1939 [1]. General interest in electromigration methods was not stimulated until Durrum's paper [2] appeared in 1950; in this he showed separations of proteins, peptides and amino acid mixtures.

At the time paper chromatography had become well known and chemists were thus used to the kind of technique involved. In the same year Kraus and Smith [3] used paper electrophoresis to determine the change of charge on the chloromercurate complex with a change of the chloride concentration in the electrolyte. Dr. Kraus told me later that he was looking for a method for the study of metal complexes in solution; after trying electromigration on paper, he decided that ion exchange chromatography was more suited for his purpose and abandoned electromigration methods.

Another pioneer of electromigration methods was H. H. Strain. In 1939 he devised a technique for column chromatography in an electric field by placing an electrode on top and another on the bottom of the adsorbent bed in a column [4]. He called the technique 'electrochromatography'. In later work he used paper strips and paper sheets but still adhered to this term.

LOW VOLTAGE PAPER ELECTROPHORESIS

Techniques

Durrum [2] let paper strips hang in a closed jar as shown in Figure 1. This set-up works well when very low currents are passed but even there the heat due to the resistance, i.e. the Joule heat, evaporates water continuously from the paper strip.

Kunkel and Tiselius [6] therefore proposed sandwiching the paper between thick glass plates: see Figure 2. Here evaporation is diminished but the

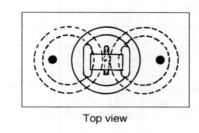

Top view

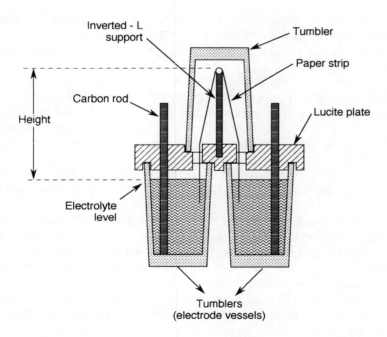

Figure 1. Diagram of Durrum's paper electrophoresis apparatus. (Reprinted with permission from Durrum [2]. Copyright (1950) American Chemical Society)

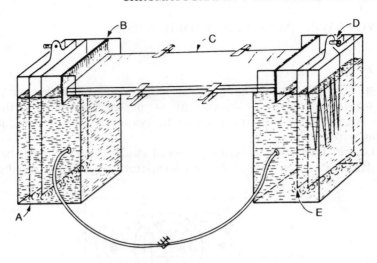

Figure 2. Apparatus of Kunkel and Tiselius [6]. A: Perspex electrode vessels; B: porous paper for establishing contact between paper and electrode vessel; C: glass plates holding the paper; D: electrodes; E: baffle plate to avoid pH changes near paper

paper still heats up to about 50–60 °C during an hour or so. The electromigration observed is thus effected in a temperature gradient.

There are also other factors to be considered in paper electrophoresis, as follows.

(i) Electro-osmosis, the global movements of the liquid in the paper due to the charges on the paper surface (mainly due to the carboxylic groups in the cellulose). This can be observed and corrected for by placing a neutral substance as indicator next to the sample. Usually glucose or hydrogen peroxide is employed (the latter, however, ionizes above pH 7, forming the peroxide anion).

(ii) Adsorption on cellulose. This can be corrected for with a chromatogram run in the background electrolyte (in a separate experiment, of course).

(iii) A tortuosity factor due to the fibre network of the paper. This will not permit a linear movement for an ion, rather a random zig-zag path as the ions move around the maze of fibres. This is best corrected for by running a substance of known mobility on the same sheet next to the sample and expressing the movement relative to this standard.

When these corrections are made one can obtain the same values for the isoelectric points of proteins as in a high precision Tiselius apparatus [6].

Below are some data for electromigration in a glass plate set-up giving relative values for electrophoretic movement in 0.42 N HBr and in acidified 0.5M SO_4^{2-} solutions. (Tables 1 and 2) They give a good picture of the actual solution chemistry in a concentration which one would use during a chemical analysis, or similar to that of sea water or serum.

Table 1. Movement of metal ions (mm) in 0.42 N HBr in one hour at 150 V using a simple glass plate apparatus [5]. (Reproduced by permission of Elsevier Science Publishers BV)

Cu^{2+}	−27	Pt^{4+}	−42	Tl^{3+}	−20
Pb^{2+}	−24	Pd^{2+}	−53	Zn^{2+}	+21
Cd^{2+}	−41	Au^{3+}	−15	Co^{2+}	+25
Bi^{3+}	−46	As^{3+}	− 4	Ni^{2+}	+23
Hg^{2+}	−51	Sb^{3+}	−17	Fe^{3+}	+23
		Sn^{2+}	− 3	Al^{3+}	+23

Table 2. Distances moved by metal ions in 0.5 M sodium sulphate solution adjusted to pH 1 with sulphuric acid in 1 h at 360 V [7]. (Reproduced by permission of Elsevier Science Publishers BV)

Metal ion	Distance moved (mm)	Metal ion	Distance moved (mm)
Ag^{I}	39	In^{III}	−14
Tl^{I}	50	Ga^{III}	− 6
Fe^{III}	21	Bi^{III}	−12
Co^{II}	22	$Co(NH_3)_6^{3+}$	12
Ni^{II}	21	Sc^{III}	−14
Zn^{II}	23	Y^{III}	9
Cd^{II}	19	La^{III}	− 6
Cu^{II}	23	Th^{IV}	−30
Mn^{II}	13	Zr^{IV}	−19
Mg^{II}	26	Ti^{IV}	−24
Be^{II}	7	V^{IV}	−11
Pb^{II}	0 (precipitated)	Ge^{IV}	− 4
Hg^{II}	0	CrO_4^{2-}	−70
Fe^{III}	−9	MoO_4^{2-}	−30
Al^{III}	9	UO_2^{2+}	−44
Cr^{III}	4		

The 'free ion', i.e. the hydrated cationic metal ion, is not so common. In the case of 0.42 N HBr the anionic species are usually reversible equilibria of series of complexes such as

$$Cd^{2+} \rightleftharpoons CdBr^+ \rightleftharpoons CdBr_2^0 \rightleftharpoons CdBr_3^- \rightleftharpoons CdBr_4^{2-}$$

and the movement indicates the average charge of the metal ion at the given anion concentration. Cd^{2+}, by the way, is cationic in 0.1 N HCl, neutral in 0.5 N HCl and anionic in 1 N HCl.

In sulphate solution there exist not only sulphato complexes but a great tendency to the formation of outer-sphere complexes also called ion pairs, between fully hydrated cations such as $Al(H_2O)_6^{3+}$ and sulphate anions.

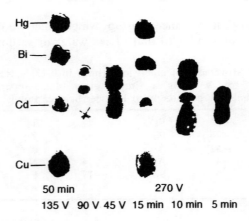

Figure 3. Separation of Hg, Bi, Cd and Cu, using 0.5 M HCl as electrolyte, by low voltage electrophoresis in a simple glass plate set-up [8]. (Reproduced by permission of Elsevier Science Publishers BV)

Figure 3 shows the separation of Hg(II), Bi(III), Cd(II) and Cu(II) in 0.5 N HCl with different times and voltages.

CONTINUOUS ELECTROPHORESIS

There was considerable interest in continuous electrophoresis during 1950–1960. Some separations from Strain and Sullivan [9] and others are shown in Figures 4–10.

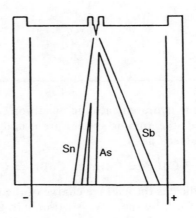

Figure 4. Separation of stannous, arsenious and antimonious chlorides [9]. Electrolyte, 0.02 M lactic acid, 0.02 M tartaric acid and 0.04 M DL-alanine; reagent, H_2S; 300 V, 95 mA

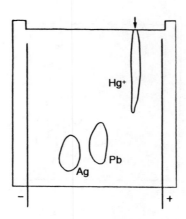

Figure 5. Discontinuous separation of mercurous, lead and silver nitrates, each 0.05 M in 1 M HNO_3 (0.01 ml). Electrolyte, 0.1 M lactic acid (60 ml); reagent, H_2S; 250 V, 100 mA, 20 min

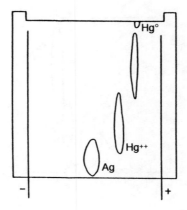

Figure 6. Discontinuous separation of mercurous, lead and silver nitrates, each 0.05 M in 1 M HNO_3 (0.01 ml). Electrolyte, 0.008 M citric acid in 4 M NH_4OH (60 ml); reagent, H_2S; 160–170 V, 100 mA, 20 min

The simple apparatus with a paper curtain has relatively low output for continuous separations, several hundred miligrams per week. Furthermore, the paths alter very much if the temperature varies during a run; some 50% mobility changes for a 10 °C change in temparature.

An improved model was constructed as shown in Figure 11. Here the liquid curtain is simply held between two glass plates. This technique found good use in complex chemistry, notably for the separation of ruthenium species. See pages 187–193.

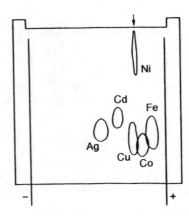

Figure 7. Discontinuous separation of nickel, ferric, cobalt, copper, cadmium and silver nitrates, each 0.05 M in 0.1 M tartaric acid (0.01 ml). Electrolyte, 0.01 M ammonium tartrate, ≈ 0.005 M dimethylglyoxime in 4 M NH_4OH (60 ml); reagent, H_2S; 160 V, 95–100 mA, 20 min

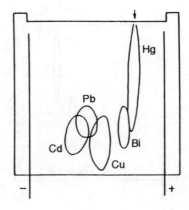

Figure 8. Discontinuous separation of mercuric, bismuth, copper, lead and cadmium nitrates, each 0.05 M in 1 M HNO_3 (0.01 ml). Electrolyte, 0.1 M lactic acid (60 ml); reagent, diphenylcarbazide (Hg), diphenylthiocarbazone (Cd), dithio-oxamide (Cu), dithio-oxamide plus NH_4OH (Pb), Na_2S (Bi); 250 V, 100 mA, 20 min

HIGH VOLTAGE PAPER ELECTROPHORESIS

The efficiency in electrophoretic separations depends on speed. As there is no equilibrium process between two phases, the faster a separation the less zone spreading due to diffusion will occur. High potentials can however only be applied if suitable cooling devices are used.

The first approach was made by Michl (for a review see Ref. 10), who immersed the paper strip in a cooling liquid, usually chlorobenzene. The

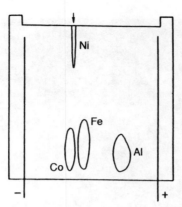

Figure 9. Discontinuous separation of nickel, cobalt, ferric and aluminium nitrates, each metal 0.005 M in 0.01 M tartaric acid and 0.005 M dimethylglyoxime (0.025 ml). Electrolyte, 0.01 M tartaric acid and 0.005 M dimethylglyoxime in 4 M NH$_4$OH (60 ml); reagent, dithio-oxamide, aluminon in 50% acetic acid; 150 V, 100 mA, 20 min

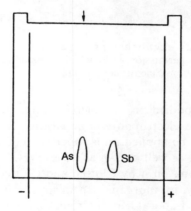

Figure 10. Discontinuous separation of arsenious and antimonious chlorides, each 0.01 M (0.05 ml). Solution and electrolyte, 0.4 M DL-alanine in 0.1 M lactic acid; reagent, H$_2$S; 300 V, 100 mA; 20 min

method is used extensively in the nucleic acid field but is questionable when solvent-soluble complexes, for example chlorauric acid, are to be examined.

Gross [11] proposed a very elegant solution to the problem of cooling. The paper strip is placed on a cooling table in which cooled brine is circulated. Complete adherence (air bubbles must be avoided!) of the paper to the cooling surface is effected via a plastic bag which is put under pressure with a bicycle pump. See Figure 12.

Figure 11(a). Continuous electrophoresis apparatus with a liquid (supportless) curtain. (Reproduced by permission of Consiglio Nazionale delle Ricerche, Ufficio Pubblicazioni e Informazioni Scientifiche, Rome)

This principle is embodied in commercial apparatus, for example the CAMAG high voltage electrophoresis apparatus. Several thousand volts, i.e. 50V/cm or so, can be applied on a paper strip 30 cm long.

The technique is not as well known as it deserves. There seems to be only one comprehensive book on it by Clotten and Clotten [12]. It has however been used extensively in inorganic chemistry, notably by Blasius and co-workers and at the Laboratorio di Cromatografia (Rome).

Some applications

Measurement of instability constants and pK values We have already mentioned above that Kunkel and Tiselius [6] obtained quite precise values for isoelectric points of proteins.

Waldmann-Meyer [13] showed that the technique also furnishes protein-binding constants, i.e. stability constants for metal–protein complexes. Values comparing several techniques are given in Table 3. Jokl and co-workers [14–16] used paper electrophoresis for determining the instability constants and ionization constants of metal ions and of condensed phosphates. Some typical examples from Jokl's work are shown in Figures 13–15.

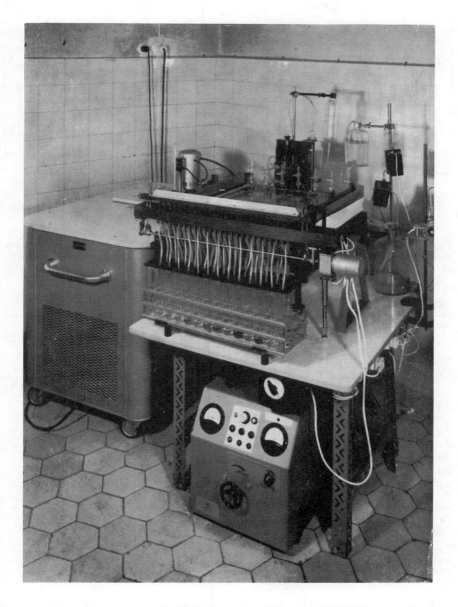

Figure 11(b). View of the supportless separating chamber. (Reproduced by permission of Consiglio Nazionale delle Ricerche, Ufficio Pubblicazioni e Informazioni Scientifiche, Rome)

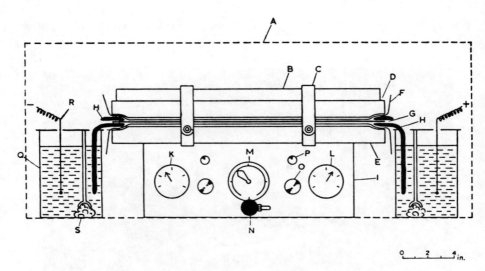

Figure 12. Apparatus according to Gross. A: safety cage, B: pneumatic pressure device, D: top cooling plate, E: bottom cooling plate, F: insulating film (polyethylene), G: paper strip, H: thick paper pad. (Reproduced by permission of CNR, Rome)

Table 3. Intrinsic cadmium and zinc binding constants [13]

Ion	Bound to	Log k_{M°	$\Gamma/2$	Medium	T, °C	Method
Cd^{2+}	4-IM^a	2.80	0.15	NaNO$_3$	25	Potentiometry
Cd^{2+}	BSA	2.80	0.15	KCl	25	Polarography
Cd^{2+}	BSA	3.84	0.20	NaAc	30	(Combined)^c
Cd^{2+}	BSA	3.12	0.10	NaAc	25	Zone electrophoresis
Zn^{2+}	4-Im^a	2.58	0.16	NaNO$_3$	24	Potentiometry
Zn^{2+}	4-Im^a	2.76	0.16	NaNO$_3$	4.5	Potentiometry
Zn^{2+}	BSA	3.08	0.05	NaAc	26	(Combined)^d
Zn^{2+}	BSA	2.9 ± 0.1	0.15	KCl	25	Polarography
Zn^{2+}	HSA	2.82 ± 0.1	0.15	NaNO$_3$	0	Equilibrium dialysis
Zn^{2+}	BSA	3.87	0.20	NaAc	30	(Combined)^c
Zn^{2+}	HSA	2.76	0.15	NaAc	25	Zone electrophoresis
Zn^{2+}	HSA	2.54	0.28	NaAc	25	Zone electrophoresis

^{a} 4-Imidazole.
^{b} Same results as with mercaptalbumin.
^{c} Equilibrium dialysis MB electrophoresis and polarography.
^{d} Polarography and equilibrium dialysis.
BSA = bovine serum albumin.
HSA = human serum albumin.

Polythionates Blasius [17] separated polythionates and similar series by paper electrophoresis and these are discussed in the chapter on sulphur, pages 170–175.

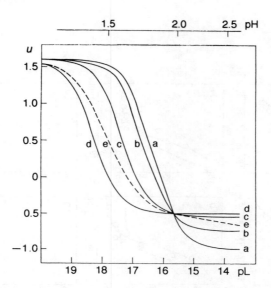

Figure 13. Calculated electrophoretic mobility curves for the Co(II)-EDTA chelate in the ligand buffers (a) pH 5; (b) pH 3; (c) pH 2 and (d) pH 1; the dotted line (e) is for a pH-buffered solution: $c_L = 10^{-2}$ M; $u_{Co} = +1.60$; $u_{CoHL} = -0.50$; $u_{CoL} = -1.00$ [14]. (Reproduced by permission of Elsevier Science Publishers BV)

Mixed ligand complexes Formation of many metal–anion complexes is instantly reversible, as exemplified by the chloride or bromide complexes of Fe(III), Co(II), Cu(II) or Cd(II). However, others such as Cr(III), Rh(III), Ru(III) etc. require some time for equilibration, and the various complexes present in a solution are readily separated by electrophoresis.

Figures 36 and 37 in Chapter 12 (vii) show the electropherograms of chloroaquorhodium (III) complexes present in various concentrations of HCl. The solutions were aged until equilibrium was established.

Separations of mixed halo-complexes such as chlorobromoiridium (IV) complexes could also be effected by paper electrophoresis.

Outer sphere complexes Metal ions with a completely filled first coordination sphere may still form a second sphere of coordination. This was first discovered by Werner, who noted that the colour (and spectrum) of the very stable $Co(NH_3)_6^{3+}$ changed in the presence of sulphate ions. In the various systems used for separation, the most striking outer sphere effects were noted in the paper electrophoresis of Co(III) complexes, as best illustrated in Figure 16 [18].

Chloride and other monovalent hydrated anions, such as acetate and nitrate, will form a non-specific anion cloud, which will retard all complexes considerably but to a similar degree. Sulphate, chromate and most divalent

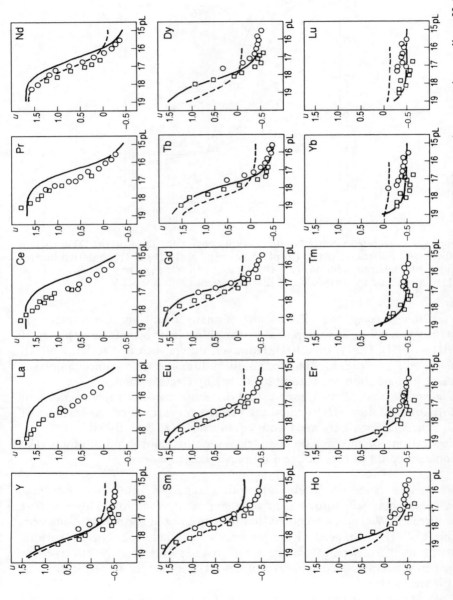

Figure 14. Calculated mobility curves and experimental data points for rare earth-EDTA complexes: continuous line pH 4; broken line pH 2; circles pH 4; squares pH 2 [15]. (Reproduced by permission of Elsevier Science Publishers BV)

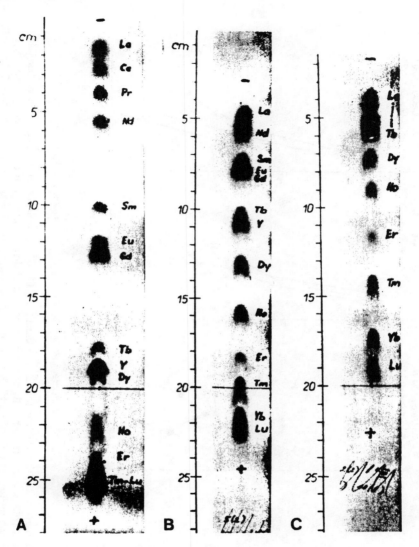

Figure 15. Paper electropherograms of the rare earths in EDTA buffers with different pL values: (A) pL 17.5; (B) pL 18.1; (C) pL 18.5 [16]. (Reproduced by permission of Elsevier Science Publishers BV)

anions will form hydrogen bridges to coordinated NH_3 or aquo groups, but not with groups such as bipyridyl or *o*-phenanthroline, which have no possibility for hydrogen bonding. This effect can be of such magnitude as to make trivalent cations behave like neutral species. Hydrophobic anions, such as perchlorate or trichloroacetate, will form 'hydrophobic ion pairs',

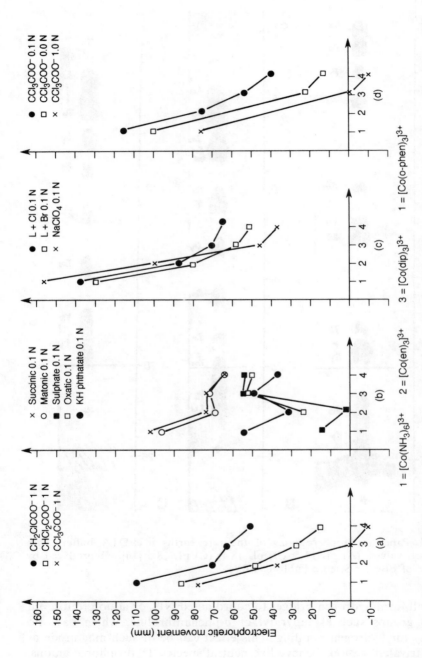

Figure 16. Graphical representation of the electrophoretic movement of Co(III) complexes in various electrolytes [18]. A, Comparison of mono-, di- and trichloroacetate (1 N); B, comparison of several divalent anions (0.1 N); C, comparison of LiCl, LiBr and NaClO$_4$ (0.1 N); D, comparison of 0.1, 0.5 and 1 N trichloroacetate. Complexes: 1 = hexamminecobalt(III); 2 = tris(ethylenediamine)cobalt(III); 3 = tris(bipyridyl)cobalt(III); 4 = tris(o-phenanthroline)cobalt(III). (Reproduced by permission of Elsevier Science Publishers BV)

especially with voluminous hydrophobic coordination groups such as bipyridyl or o-phenanthroline. Again, the effect can be very strong.

In ion exchange, outer sphere complexation between the sulphonic groups of a resin and the aquo or ammine groups of a complex has been shown to exist. It should be stressed here that inner complex formation has also been observed. There is also evidence that outer sphere complexation between, e.g., aluminate groups on the surface of alumina or silanol groups on the surface of silica and coordination complexes does play a part in the chromatography of such complexes on inorganic exchangers.

In most techniques only a change in mobility or of a two-phase equilibrium is observed, which does not indicate whether inner sphere or outer sphere complexes are formed. In the case of the Co(III) complexes, which are very stable indeed, the inner sphere complexes are ruled out. With hexa-aquo Al(III) or hexa-aquo Fe(III), other evidence is needed for establishing the kind of complex formed.

CROSS ELECTROPHORESIS

It is possible to let an anion migrate through a zone of a cation in electrophoresis. If they do not interact they will cross without either altering the movement of the other. If they do interact, they may change colour or mobility during the time that they are in contact.

The first observations were reported in 1953 by Grassmann and Hübner using an apparatus for continuous electrophoresis. The same kind of information can also be obtained on a paper strip, for example by allowing a small round spot migrate through a wider thin band. Typical patterns for enzyme–substrate interactions are shown in Figure 17.

Care must be taken that the crossing band has the same conductivity as the spot. Otherwise a deformation may be observed simply due to a change in conductivity in the absence of an interaction.

The technique has been widely used in studying protein interactions. A book by Nakamura on this subject is available [19]. There is little work on inorganic reactions [20]: when a band of Fe(III) migrates against a spot of CNS^- there is a colour change during their crossing but no zone deformation. It seems that the reaction $FeCl^{2+} \rightleftharpoons FeCNS^{2+}$ occurs without a change in mobility.

GEL ELECTROPHORESIS

Consden et al. [21] separated partial hydrolysates of wool in 100 cm long slabs of silica gel. Later authors preferred agar or agarose gel, as it does not adsorb proteins as silica gel does. Inorganic ions may be separated in agar gel [22] as well as in starch gels [23]. However, these gels are not compatible with strong acids as electrolytes and thus buffers or neutral salts must be used.

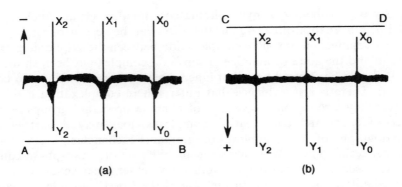

Figure 17. Cross electrophoresis of active arginase with arginine. (a) AB: 1.43 µl/cm 0.01 M arginine solution. X_0Y_0, X_1Y_1, X_2Y_2: µl/cm 0.5% solution of active arginase, inactive (dialysed against phosphate buffer) and inactive (dialysed against phosphate buffer containing 0.55% EDTA) arginase, respectively. At 250 V, 0–3 °C for 90 min; phosphate buffer, pH 6.8, ionic strength 0.05; dried at 110 °C and stained with ninhydrin. (b) Control: CD, 0.7 µl 0.5 mM glutamate solution. (Reproduced by permission of Elsevier Science Publishers BV)

Rather exciting preparative separations by gel electrophoresis were described by Blasius and co-workers [24, 25], who used a PVC elastic tube filled with the gel (cellulose acetate gel) and organic solvents both for isotachophoresis and zone electrophoresis. The separated zones were recovered by cutting up the gel tube.

Figure 18 shows the apparatus. The tubes are 1 metre long or more and runs at −4°C are performed for 50 hours or so at current strengths of 4–5 mA with solvents such as acetonitrile or acetonitrile–alcohol (1:1). A typical separation is shown schematically in Figure 19.

ELECTROPHORESIS IN FUSED SALTS

A considerable amount of work has been published on electromigration of inorganic ions in fused salts. For a review see [26]. As support, glass fibre

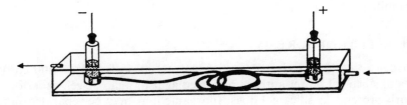

Figure 18. Apparatus for preparative electrophoresis in a PVC tube using organic solvents [24]. (Reproduced by permission of Elsevier Science Publishers BV)

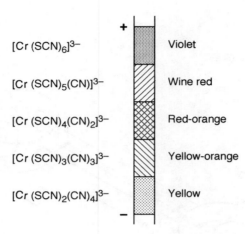

Figure 19. Separation of mixed cyano–thiocyanato Cr(III) complexes by isotachophoresis in a gel with organic solvent [25]. (Reproduced by permission of Elsevier Science Publishers BV)

paper, asbestos papers or sheets, ceramic thin layers and also columns filled with sand or quartz powder have been used.

Quite a wide range of eutectics can be made which are liquid at 160–250°C. The interest in this work was in isotope effects and in the measurement of mobilities for the elucidation of complex equilibria in fused salts, thus with non-hydrated ions.

Typical electrophoretic apparatus is shown in Figures 20 and 21. The electrode vessels and the support for the glass fibre paper are made in one piece of vitreous silica. Both a furnace for constant temperature and a controlled atmosphere are provided.

Typical mobilities and some representative separations are shown in Table 4 and Figures 22 and 23.

Table 4 Movement of metal ions in fused salts

	Distance (cm) moved in 4 h	KCl—LiCl eutectic $t = 450°$	KNO—LiNO$_3$ eutectic $t = 160°$; 5 V/cm
Anionic	3 −0.5	Zn(II), Co(II)	
Isoelectric		Th(IV)^a	Th(IV)^a
	0.5–3	Ce(III)	Cd(II)
	3 −5.5	Pb(II), Cd(II)	Pb(II)
Cationic	5.5–8	Cu(II)	Sr(II), Ba(II), Cs(I), Rb (I)
	8 −10.5	Cs(I), Rb(I)	
	10.5–13	Na(I), Ag(I)	

^{a} Insoluble precipitate formed.

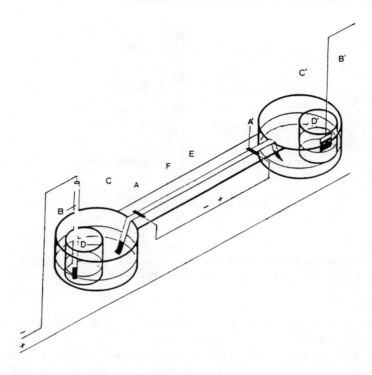

Figure 20. Apparatus for zone electrophoresis in molten salts. A, A': platinum wires for measurements of potential difference; B, B': electrodes; C, C': reservoirs; D, D': electrode compartments provided with sintered glass discs at the bottom; E: supporting glass plate; F: electrophoretic strip [2]

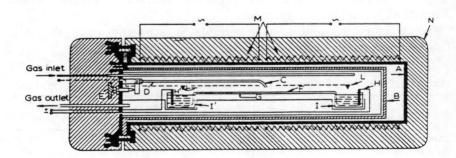

Figure 21. Cross-sectional view of an electrophoretic apparatus. A: furnace; B: electrophoresis chamber; C: capillary; D: tube supporting the capillary; E: screw to raise or lower tube D; F: glass fibre paper; G: Pyrex glass plate; H, H': graphite electrodes; I, I': Pyrex vessels; L, L': thermocouples; M: Ni–Cr heating wire; N: insulating jacket [2]. (Reproduced by permission of Elsevier Science Publishers BV)

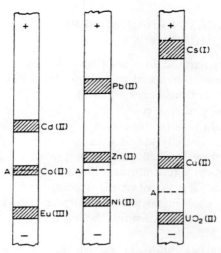

Figure 22. Some representative separations of inorganic ions by zone electrophoresis in molten $LiNO_3$–KNO_3 eutectic- 10% NH_4NO_3 at 100 °C. A: application point [27]

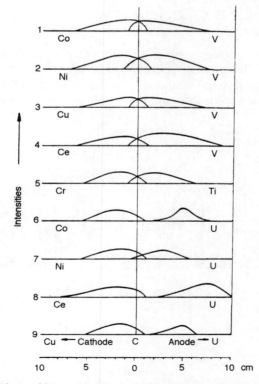

Figure 23. Separations of inorganic ions by column electrophoresis in molten $KHSO_4$–$K_2S_2O_7$ eutectic [28]. (Reproduced by permission of Elsevier Science Publishers BV)

ISOELECTRIC FOCUSING (FOKUSSIERENDER IONENAUSTAUSCH)

This is an ingenious method first used by Schumacher [29]) for inorganic ions and later extensively worked out for proteins (see Ref. 30). A rather short paper strip (glass fibre paper and PVC and cellulose ester paper have also been used) is dipped in a complexing cathode solution and a complex destroying anode solution. In the centre is a fairly wide zone of the metal ions to be separated. Usually the entire arrangement is dipped in a cooling solution such as carbon tetrachloride. A very high voltage, something like 100 V/cm, is applied for a few minutes. During this time, if conditions have been correctly chosen, a gradient of pH produces a gradient of complexing conditions in which each metal will concentrate as a thin band depending on its complexing stability constant. Below are some examples and the apparatus (Figures 24 and 25).

Except for Schumacher and his group this method has not attracted much interest.

ISOTACHOPHORESIS OR DISPLACEMENT ELECTROPHORESIS

In chromatography (irrespective of the mechanism), an arrangement can be set up to produce displacement effects. Let us take the case of an adsorption column on which several substances are adsorbed together at the top as a single band. If we wash this column with a solution of a substance which

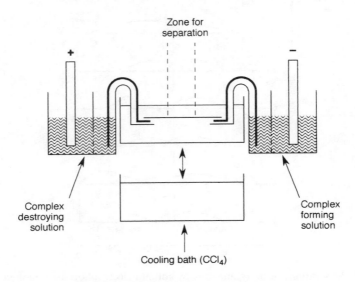

Figure 24. Apparatus for Schumacher's isoelectric focusing

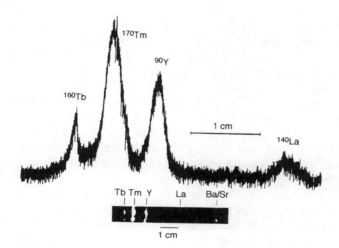

Figure 25. Radioactive scan and radioautogram of a separation of rare earths with 0.5 M HCl and 0.3 M complexing solution for 8 min at 500 V

is more strongly adsorbed than the constituents of the mixture placed on the column, these adsorbed substances will be pushed down the column. If the desorption–adsorption equilibria are repeated a sufficient number of times, then there will be a sequence of separated substances moving down the column, with the least adsorbed eluting first followed by the others in order of their strengths of adsorption and each being followed immediately by the next. There will be no empty zones on the adsorbent between them.

Electrophoretic methods usually have analogues to the chromatographic process. Zone electrophoresis is analogous to elution chromatography. The Tiselius U-tube technique is analogous to frontal analysis. The analogue to displacement development is isotachophoresis. Martin toyed with ideas how to effect this for quite some time (from 1942 onwards), and when he was 'extra-mural professor' at Eindhoven he worked it out together with Everaerts [31].

The principle can be applied in any type of apparatus. Isotachophoresis has been performed on paper, on gel slabs, in tubes filled with gels, in 'free electrophoresis' in capillary tubes (this being the preferred analytical technique), and in continuous electrophoresis with or without a stabilizing support (the latter being the preferred preparative technique).

The procedure is as simple as zone electrophoresis if you have some data on the mobilities of the ions to be separated. You place your mixture as a short or even a longer zone into your set-up. It must be preceded by a faster ion called the 'leading electrolyte' and behind it must be a slower ion called the 'terminating electrolyte'. You then get the whole train to migrate by applying a voltage compatible with the support and the apparatus. When 'equilibrium conditions' are reached the leading electrolyte is followed by the fastest ion of the mixture to be separated, then the next fastest and so on until, at the rear, comes the terminating electroyle. In this process all zones are either concentrated or diluted so that they all have the same migration velocity. It is thus possible to concentrate traces during a separation.

What if ions in the mixture are faster than the leading or slower than the terminating ion? Then they will migrate in the respective electrolyte without compression into the isotachophoretic train and without concentration or dilution effects.

Students often have difficulties with this principle and hence another explanation is offered, this time that by Everaerts [31].

We shall consider here the separation of anionic species in narrow-bore tubes. For the separation of anionic species, the narrow-bore tube and anode compartment are filled with the so-called leading electrolyte, the anions of which must have a mobility that is higher than that of any of the sample anionic species. The cations of the leading electrolyte must have a buffering capacity at the pH at which the analyses will be performed. The cathode compartment is filled with the terminating electrolyte, the anions of which must have a mobility that is lower than that of any of the sample anionic species. The sample is introduced between the leading and terminating electrolyte, e.g., by means of a sample tap or a micro-syringe.

When an electric current is passed through such a system (see Figure 26(a)), a uniform electric field strength over the sample zone occurs and hence each sample anionic species will have a different migration velocity. The sample anionic species with the highest effective mobility will run forwards and those with lower mobilities will remain behind.* Hence, both in front of and behind the original sample zone, the moving-boundary procedure results in two series of mixed zones (comparable with the Tiselius method). In the series of mixed zones, the sample anionic species are arranged in order of their decreasing effective mobilities (see Figure 26(b)).

The anionic species of the leading electrolyte can never be passed by sample anions, because its effective mobility is chosen so as to be higher. Similarly, the terminating anions can never pass the anionic species of the sample. In this way, the sample zones are sandwiched between the leading and terminating electrolyte. In the mixed zones of the sample, the separation continues and, after some time, when the separation is complete, a series of zones is obtained in which each zone contains only one anionic

---

*If it is assumed that the zones release, then the concentration of the ionic species at that position will decrease, the electric field strength will increase (working at a constant current density) and hence the migration velocity of the ionic species involved will be higher. Therefore, finally these ionic species will reach the preceding zone.

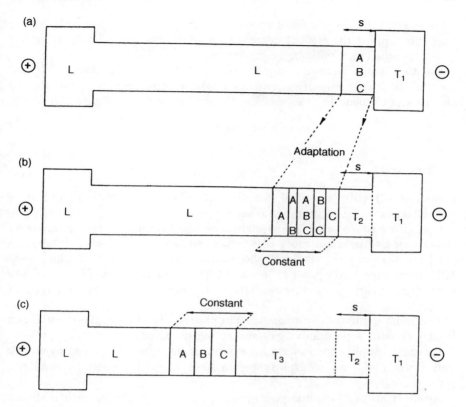

Figure 26. Separation of a mixture of anions according to the isotachophoretic principle. The sample A + B + C is introduced between the leading anionic species L and the terminating anionic species T. A suitable cationic species is chosen as the buffering counter-ion. The original conditions are shown in (a). After some time (b), some mixed zones are obtained according to the moving boundary principle. Finally (c), all anionic species of the sample are separated and all zones contain only one anionic species of the sample ('ideal case')

species of the sample if no anionic species with identical effective mobilities are present in the sample. Of course, this series of zones is still sandwiched between leading and terminating electrolyte (see Figure 26(c)).

The first sample zone contains the anionic species of the sample with the highest effective mobility, the last zone that with the lowest effective mobility. After this stage, no further changes to the system occur and a steady state has been reached. In such a case, we can speak of an isotachophoretic separated system. (Of course, one or more unmixable 'mixed zones', i.e., zones that contain one or more anionic species with identical effective mobilities, may still be present.) In this state, all of the zones must run connected together, in contrast to zone electrophoresis, where all zones release. Here the zones cannot release as there is no background electrolyte that can support the electric current (a requirement for the solvent is that its self-conductance must be negligible).

For this steady state, all zones must have identical migration velocities, determined by the migration velocity of the anionic species of the leading electrolyte. Considering the zones L, A, B, C and T (see Figure 26(c))

$$v_L = v_A = v_B = v_C = v_T \tag{1}$$

or

$$m_L E_L = m_A E_A = m_B E_B = m_C E_C = m_T E_T \tag{2}$$

Equation (2) will be called the 'isotachophoretic condition' and it is characteristic of isotachophoretic separations.

It is also easy to see what happens when isotachophoresis is effected on paper strips with one or more coloured substances [32, 33]. Incidentally, no further work on paper isotachophoresis has been published. Taglia joined IBM to become a computer expert and I had other interests. There is plenty of scope for further work, especially if the technique is used as a type of spot test.

The most exploited form of isotachophoresis is that in glass capillaries linked up with a conductivity or spectrometric detector.

Inorganic applications of isotachophoresis have been reviewed by Boček and Foret [34], and several symposia have been dedicated to isotachophoresis.

An illustration of the analytical possibilities of isotachophoresis is shown in Figure 27.

For the separation of ruthenium nitrosylnitrato complexes see page 193.

CAPILLARY ZONE ELECTROPHORESIS

During work on capillary isotachophoresis, Everaerts et al. [31] made the important observation that zone electrophoresis was equally feasible in capillaries. This was not really new, as it had been worked out for somewhat wider thin tubes (3 mm diameter) by Hjertèn in 1967 [35]. The technique became popular, however, only when work with capillaries had become widespread in connection with isotachophoresis.

This is a good moment to reflect on the various electrophoretic techniques. The Tiselius apparatus was the result of research towards an instrument which permitted precise measurement. It was not popular with analytical chemists who, in the 1940's, were not used to working with complicated apparatus; nor was it usual to spend funds on such apparatus in analytical work.

When paper electrophoresis became popular, chemists were accustomed to spot tests and paper chromatography, so that this new technique did

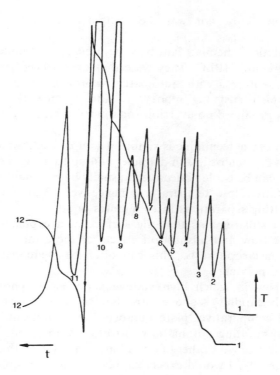

Figure 27. Isotachopherogram of the separation of some anions in the operational system listed below [31]. 1 = Chloride; 2 = nitrate; 3 = oxalate; 4 = tartronate; 5 = formate; 6 = citrate; 7 = maleate; 8 = adipate; 9 = iodate; 10 = trichloroacetate; 11 = mandelate; 12 = ascorbate. The current was stabilized at 70 µA. T = Increasing temperature; t = time. (Reproduced by permission of Elsevier Science Publishers BV)

Operational system at pH 6 suitable for anionic separations

Solvent: H_2O and D_2O
Electric current: ≈50–100 µA
Purification: Morpholinoethanesulphonic acid (MES) is recrystallized three times and the crystals are washed with acetone

	Electrolyte	
	Leading	Terminating
Anion	Cl^-	E.g., MES^-
Concentration	0.01 N	≈ 0.01N
Counter-ion	Histidine	$Tris^+$
pH	6.02	≈ 6
Additive	0.05% polyvinyl alcohol (Mowiol)[a]	None

[a] For experiments with a thermometric detector, this additive is not necessary.

not require new skills, nor was the equipment (power pack etc.) very expensive.

By 1979 analytical chemists had become used to techniques such as gas chromatography and HPLC. They were thus experienced in working with still smaller quantities, with automatic recording equipment and with instruments which cost big money. In this framework capillary zone electrophoresis produced no aversions or shocks, which it would have done in 1960.

The electrophoretic technique is simple: a sample is allowed to migrate in long capillaries 0.2 mm or less in diameter and up to many metres in length. The capillaries can be cooled effectively, usually by placing them in a beaker at a thermostatted low temperature; hence very high voltages can be applied, permitting separations in 10 minutes or so.

The problem still remains that the surface in a capillary can provoke electro-osmotic flow and adsorption effects of the same order as in high voltage paper electrophoresis. This can often be overcome by lining the capillary with a polymer.

The few papers in which high voltage paper electrophoresis has been compared with capillary zone electrophoresis seem to show that both methods yield comparative 'plate numbers' or resolutions. However, the apparatus incorporating automatic registration with an ultra-sensitive detector is more in line with other 'modern' instrumentation than the one with long paper strips. So I would expect capillary zone electrophoresis to have a bright future.

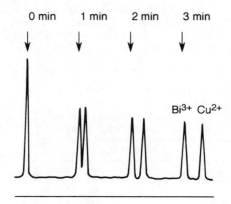

Figure 28. Separation of inorganic ions (Cu^{2+} and Bi^{3+}) by free zone electrophoresis [35]. Medium, 0.1 M lactic acid. Sample: 10 µg of $Cu(NO_3)_2 \cdot 3H_2O$ and 7 µg of $Bi(NO_3)_3$ in 8 µl of 0.1 M lactic acid. Inner diameter of electrophoresis tube, 3 mm; current, 8 mA; voltage, 2540 V; temperature of cooling water, 10 °C. The arrow indicates the position of the starting zone. The scans were made at the times indicated. (Reproduced by permission of Elsevier Science Publishers BV)

Now some examples of separations. Figure 28 shows the separation of Bi(III)–Cu(II) in 3 minutes by using Hjertèn's apparatus. Figure 29 depicts the separation of 15 cations in 8 minutes, from an early paper by Weston et al. [36].

HIGH PERFORMANCE PAPER ELECTROPHORESIS

In 1979, when the development of sophisticated high performance methods was in full swing, we wanted to see whether simple apparatus could not be improved likewise.

If the scale of usual low voltage electrophoresis is reduced to paper strips 6 mm wide and 100 mm long the Joule heat generated is so much less that much higher voltages can be applied even when only glass plates are used for cooling. Thus, instead of 30–60 minutes, only 5 minutes suffice for an equivalent separation. Some pairs of ions, for example ferrocyanide–ferricyanide, will separate in 30 seconds.

Figure 30 shows some separations from the first report on this technique [37]. For further work see [38].

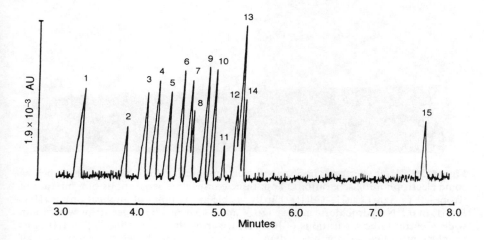

Figure 29. Capillary zone electrophoresis [36]. Separation of alkali, alkaline earth and transition metal cations with the aid of an alternative complexing agent, HIBA (α-hydroxyisobutyric acid). Carrier electrolyte, 5 mM Waters UVCat-1–6.5 mM HIBA (pH 4.4); capillary as described previously; voltage, 20 kV (positive); hydrostatic injection, as before; indirect UV detection at 214 nm. Peaks: 1 = potassium (0.8 ppm); 2 = barium (1.5 ppm); 3 = strontium (1.5 ppm); 4 = calcium (0.7 ppm); 5 = sodium (0.6 ppm); 6 = magnesium (0.4 ppm); 7 = manganese (0.8 ppm); 8 = cadmium (0.8 ppm); 9 = iron(II) (1.0 ppm); 10 = cobalt (0.8 ppm); 11 = lead (1.0 ppm); 12 = nickel (0.6 ppm; 13 = lithium (0.2 ppm); 14 = zinc (0.4 ppm); 15 = copper (0.6 ppm). (Reproduced by permission of Elsevier Science Publishers BV)

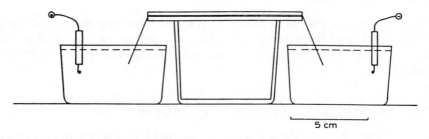

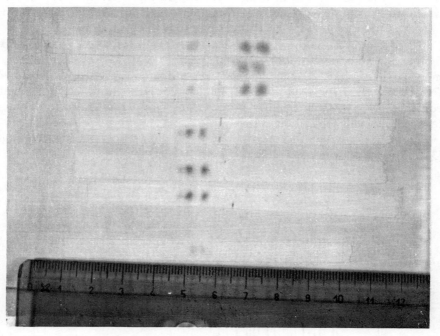

Figure 30. 'High performance' paper electrophoresis. Above: apparatus; below: some electrophoretic separations. Top: three consecutive separations of a mixture of (from left to right) Cu(II)–Cd(II) (hardly visible at the point of application)–Bi(III)–Hg(II) in 0.5 N hydrochloric acid at 180 V and ≈45 mA. The paper strips were 6 mm wide. Centre: three separations of [Co(III) trisorthophenanthroline]$^{3+}$–[Co(III) diorthophenanthroline monoethylenediamine]$^{3+}$–Co^{2+}. (Reproduced by permission of Elsevier Science Publishers BV)

References

[1] D. von Klobusitzky and P. König, *Arch. Exp. Pathol. Pharmacol.* **192**(1939) 271.
[2] E. L. Durrum, *J. Am. Chem. Soc.* **72**(1950) 2943.
[3] K. A. Kraus and G.W. Smith, *J. Am. Chem. Soc.* **72**(1950) 4329.
[4] H. H. Strain, *J. Am. Chem. Soc.* **61**(1939) 1292.
[5] M. Lederer, *An Introduction to Paper Electrophoresis*, Elsevier, Amsterdam (1955), p. 145.

[6] H. Kunkel and A. Tiselius, *J. Gen. Physiol.* **35**(1951) 89.
[7] H. C. Chakrabortty, *J. Chromatogr.* **5**(1961) 121.
[8] M. Lederer and F. L. Ward, *Anal. Chim. Acta* **6**(1952) 355.
[9] H. H. Strain and J. C. Sullivan, *Anal. Chem.* **23**(1951) 816.
[10] H. Michl, *Chromatogr. Rev.* **1**(1959) 11–38.
[11] D. Gross, *Nature* **180**(1957) 596.
[12] R. Clotten and A. Clotten, '*Hochspannungs-Elektrophorese*', Georg Thieme Verlag, Stuttgart (1962).
[13] H. Waldmann-Meyer, *Chromatogr.Rev.* **5**(1963) 1–41.
[14] J. Jokl, *J. Chromatogr.* **71**(1972) 523.
[15] J. Jokl and I. Valášková, *J. Chromatogr.* **72**(1972) 373.
[16] J. Jokl and Z. Pikulikova, *J. Chromatogr.* **74**(1972) 325.
[17] E. Blasius, K. Müller and K. Ziegler, *J. Chromatogr.* **313**(1984) 165.
[18] M. Lederer and M. Mazzei, J. Chromatogr. **35**(1968) 201.
[19] S. Nakamura, *Cross Electrophoresis*, Elsevier, Amsterdam (1966).
[20] V. Mosini and M. Lederer, *J. Chromatogr.* **77**(1973) 464.
[21] R. Consden, A. H. Gordon and A. J. P. Martin, *Biochem. J.* **40**(1946) 33.
[22] M. Lederer and I. Cook, *Aust J. Sci.* **14**(1951) 56.
[23] G. Cetini, *Ann. Chim. (Rome)* **45**(1955) 216.
[24] E. Blasius and U. Wenzel, *J. Chromatogr.* **49**(1970) 527.
[25] E. Blasius, H. Augustin and U. Wenzel, *J. Chromatogr.* **50**(1970) 319.
[26] G. Alberti and S. Alluli, *Chromatogr. Rev.* **10**(1968) 99.
[27] G. Alberti, G. Grassini and R. Trucco, *J. Electroanal. Chem.* **3**(1962) 283.
[28] H. Kühnl and M. A. Khan, *J. Chromatogr.* **23**(1966) 149.
[29] E. Schumacher, *Helv. Chim. Acta* **40**(1957) 221.
[30] P. G. Righetti, *Isoelectric Focusing*, Elsevier Biomedical Press, Amsterdam (1983), 386 pp.
[31] F. M. Everaerts, J. L. Beckers and Th. P. E. M. Verheggen, *Isotachophoresis*, Elsevier, Amsterdam (1976), 418 pp.
[32] V. Taglia and M. Lederer, *J. Chromatogr.* **77**(1973) 467.
[33] V. Taglia, *J. Chromatogr.* **79**(1973) 380.
[34] P. Boček and F. Foret, *J. Chromatogr.* **313**(1984) 189.
[35] S. Hjertèn, *Chromatogr.Rev.* **9**(1967) 122.
[36] A. Weston, P. R. Brown, P. Jandik, W. R. Jones and A. L. Heckenberg, *J. Chromatogr.* **593**(1992) 289.
[37] M. Lederer, *J. Chromatog.* **171** (1979) 403.
[38] S. Fanali and L. Ossicini, *J. Chromatog.* **212** (1981) 374.

5 GEL FILTRATION

Chromatographic and electrophoretic effects due to the *size* of a species are quite numerous. One of the early observations in inorganic chromatography is that of Ayres [1] that zirconium is eluted separately from Fe (III), Be (II) etc. because zirconium exists in dilute acid as a hydrolysis polymer too large to enter the pores of a cation exchange resin.

The preparation of special media for separations according to molecular size dates from 1959, when Porath and Flodin [2] prepared dextrans cross-linked with epichlorhydrin. A network with controlled pore size depending on the degree of cross-linking is the result, Figure 1. A schematic representation of several such gel types is shown in Figure 2.

If we assume that the pores are conical in shape it is easy to picture the separation mechanism. Firstly, all molecules too large to enter the pores will be excluded altogether; secondly, molecules of a size able to enter the pores will remain in the gel depending on how far they can penetrate. Thus, the larger the molecule the faster its elution.

Selectivity curves for a series of Sephadex G (cross-linked dextrans) with varying degree of cross-linking are shown in Figure 3. On columns of Sephadex it is possible to estimate the molecular *size* (usually identical with molecular weight within a group of compounds of similar or identical shape) of numerous groups of compounds.

It was evident that this field also interested inorganic chemists. At first sight, gel chromatography would be the ideal method for the isolation and characterization of hydrolysis polymers. However, most of the present gel filtration media are unsuitable for such work. Sephadex gels have on their surface vicinal hydroxyl groups which complex readily with most metal ions at neutral pH (a typical compound of this type is the 'glycerate of iron' of the British Pharmacopoeia). Hence, these ions leave a trail when applied in high concentrations, and they are adsorbed altogether in low concentrations. Porous glasses and porous silica (and their derivatives) usually have

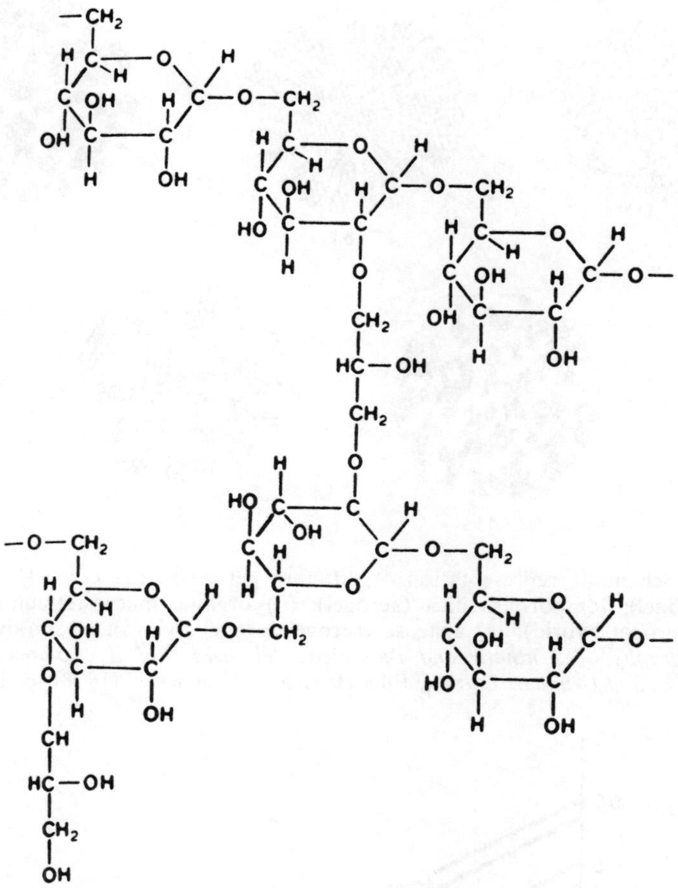

Figure 1. Schematic representation of Sephadex. (Reproduced from *Dextran Gels and Their Application in Gel Filtration*, Pharmacia, Uppsala, 1962)

on their surfaces a high concentration of silanol groups, which also interact strongly with monomeric and polymeric metal ions.

In spite of these shortcomings, quite a few polymeric hydrolytic species could be separated on gel filtration media, e.g.: polymeric ruthenium [3], polymeric rhodium (III) [4], and soluble ferrocyanides [5–7]. The movement of small monomeric ions inside gel filtration media is largely governed by ion exchange with residual carboxyl groups, ion exclusion due to these groups, or hydrophobic adsorption, which is rather strong on Sephadex LH-20, where also salting-out effects can be observed [8].

Gel chromatography has been used to study the interaction between metal ions and large molecules, such as proteins, dextrans, and polyphosphates.

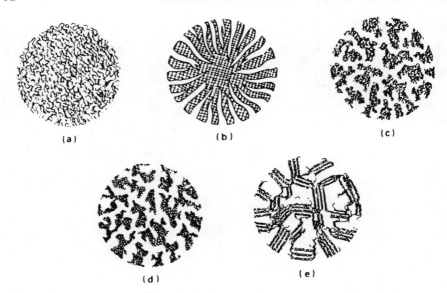

Figure 2. Schematic representation of different gel types: (a) xerogel; (b) porous glass (aerogel); (c) porous silica (aerogel); (d) organic macro-reticular polymer (xerogel–aerogel hybrid); (e) agarose (xerogel-aerogel hybrid). (Reproduced from *Chromatography of Synthetic and Biological Polymers, Vol. 1, Column Packings, G.P.C., G.F. and Gradient Elution*, Ellis Horwood, Chichester (1978), p. 1)

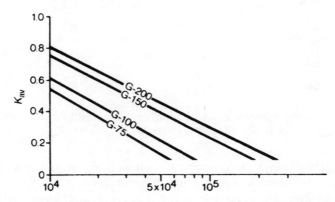

Figure 3. The distribution coefficient plotted against the molecular mass (for globular proteins) for several types of Sephadex G with different degrees of cross-linking. (From *Filtration sur Gel*, Pharmacia, Uppsala)

Most of the early work is discussed in an excellent review by Yoza [9]. Another fascinating area that can be studied by gel chromatography deals with the interactions between inorganic polymers and small ions. For instance, the zirconium hydrolysis polymer is excluded from Sephadex, but small ions, such as chromate, are not. When a mixture of polymeric zirconium and chromate is chromatographed on Sephadex G-10, there is a slow-moving yellow band as well as a yellow excluded band, showing clearly that the polymer binds some of the chromate [10].

One important problem of inorganic chemistry was approached, if not solved, by gel filtration, namely the solution chemistry of silicates. There was a recent review in this field by Tarutani [11]. It is possible to separate polysilicic acids from monosilicic acid on a Sephadex G-100 column and, although silicates are soluble only to some hundred ppm, they can be estimated colorimetrically after elution. It is thus possible, for example, to study the effect of ageing of a monosilicate solution at various pH values, see Figures 4 and 5, or the polymerization of silicic acid in geothermal waters (Figure 6).

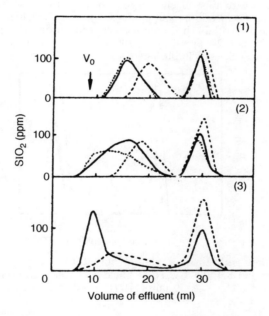

Figure 4. Variation with time of the elution curves for silicic acid in solutions of pH 9.5, 8 and 6. Initial monomer concentration: 500 ppm (SiO_2); gel: Sephadex G-100; column: 45 cm × 1.0 cm i.d. (1) pH 9.5: ----, 6 h (monomer concentration 215 ppm); ——, 100 h (192 ppm); ...., 250 h (174 ppm). (2) pH 8: ----, 2 h (230 ppm); ——, 24 h (147 ppm); ...., 75 h (136 ppm). (3) pH 6: ----, 24 h (298 ppm); ——, 75 h (186 ppm) [11]. (Reproduced by permission of Elsevier Science Publishers BV)

Apart from purely chemical and geological interest there seems also to be a medical aspect. We ingest in most parts of the world a saturated solution of silicates during our entire life. As the amount in drinking water is usually in the ppm range and difficult to determine, it is usually ignored. However,

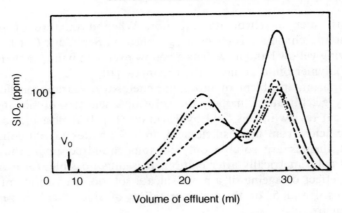

Figure 5. Variation with time of elution for silicic acid in 1 M hydrochloric acid. Initial monomer concentration: 500 ppm (SiO_2); gel: Sephadex G-100; column: 45 cm × 1.0 cm i.d. ──, 24 h (monomer concentration 367 ppm); ---, 50 h (278 ppm); ....., 75 h (227 ppm); -.-.-, 100 h (204 ppm) [11]. (Reproduced by permission of Elsevier Science Publishers BV)

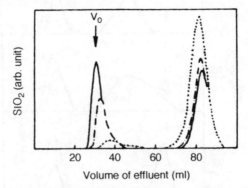

Figure 6. Variation with time of elution curves for silicic acid in Hachobaru geothermal water (pH 8.3, 90 °C). Total silicic acid concentration: 950 ppm (SiO_2). ....., 5 min; -.-.-, 15 min; ──, 30 min [11]. (Reproduced by permission of Elsevier Science Publishers BV)

we should consider its fate. Is it excreted and how? How does it react with (or complex) essential or poisonous trace metals? Silica in the form of silica gel adsorbs metal ions strongly by ion exchange, complexation and outer sphere complexation, so it is probable that the surface of polysilicates will react likewise. To my knowledge this field has not been studied as yet.

References

[1] J. A. Ayres, *J. Am. Chem. Soc.* **69**(1947) 2879.
[2] J. Porath and P. Flodin, *Nature* **183**(1959) 1657.

[3] I. Kitayevitch, M. Rona and G. Schmuckler, *Anal. Chim. Acta* **61**(1972) 277.
[4] M. Sinibaldi and A. Braconi, *J. Chromatogr.* **94**(1974) 338.
[5] H. Saito and Y. Matsumoto, *J. Chromatogr.* **168**(1979) 227.
[6] Y. Matsumoto, M. Shirai and H. Saito, *Bull. Chem. Soc. Jpn.* **41**(1975) 210.
[7] D. Corradini and M. Sinibaldi *J. Chromatogr.* **187**(1980) 458.
[8] V. DiGregorio and M. Sinibaldi, *J. Chromatogr.* **129**(1976) 407.
[9] N. Yoza, *J. Chromatogr.* **86**(1973) 325.
[10] M. Sinibaldi, G. Matricini and M. Lederer, *J. Chromatogr.* **129**(1976) 412.
[11] T. Tarutani, *J. Chromatogr.* **313**(1984) 33.

6 ION EXCHANGE

According to Rieman and Walton [1], ion exchange is mentioned in the bible, and since 1850 has been recognized as a phenomenon in soil chemistry.

A definition is difficult. Perhaps the best is that ion exchange is a stoichiometric reaction between a solid and a solution whereby an ion is taken up by the solid and another goes into solution.

The first synthesis of an ion exchange resin was carried out in 1935 by Adams and Holmes [2], who prepared a condensation product of phenol-sulphonic acid with formaldehyde. Similar substances were prepared by Liebknecht [3] and Smit [4] by sulphonation of coal. All these resins possessed reactive OH and COOH in addition to the more important SO_3H exchange groups.

In order to prepare a resin with only one type of reactive group, D'Alelio [5] sulphonated a hydrocarbon polymer containing benzene rings (styrene with 10% divinylbenzene). An analogous resin with basic groups was prepared by reacting the polymer with chloromethyl ether,

$$\text{–CH–CH}_2\text{–}\text{C}_6\text{H}_5 + CH_3OCH_2Cl \rightarrow \text{–CH–CH}_2\text{–}\text{C}_6\text{H}_4\text{CH}_2\text{Cl} + CH_3OH$$

then reacting the chloro groups in the network with tertiary amines.

$$\text{–CH–CH}_2\text{–}\text{C}_6\text{H}_4\text{CH}_2\text{Cl} + R_3N \rightarrow \text{–CH–CH}_2\text{–}\text{C}_6\text{H}_4\text{CH}_2NR_3 + Cl$$

ION EXCHANGE

The functional acidic (or basic) groups of an exchanger will always be occupied by ions of the opposite charge. When holding hydrogen ions a resin is said to be in the 'hydrogen form', similarly when holding sodium ions in the 'sodium form' etc.

When a polymer containing active groups (e.g. SO_3H) is formed, the ionization of the respective groups is not changed, thus a *vast sponge-like network* is produced with properties identical with those of the monomer. Resins with sulphonic groups or quaternary amine groups are thus highly ionized, though very insoluble, and react throughout their entirety. Resins with highly ionized groups such as SO_3H and NR_3 are called strong exchange resins, and resins with only partially ionized groups such as COOH, OH, NH_2 are called weak exchange resins. The degree of ionization as well as the similarity to the monomer can best be illustrated by titrating, for example, the hydrogen form of a strong acid resin such as Dowex-50. A titration curve identical to that of a strong acid with a strong base is produced (Figure 1).

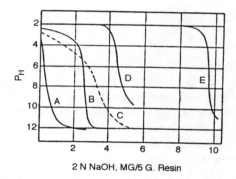

Figure 1. Titration curves of several cation exchanger functional groups; A, phenolic OH: B, methylenesulphonic CH_2SO_2H: C; carboxyl COOH: D and E, nuclear SO_3H (Tompkins [6]; reprinted with permission from *Anal. Chem.* **22** (1950) 1352. Copyright (1950) American Chemical Society)

A similar analogy exists with weak exchange resins, these giving titration curves typical of weak acids and weak bases.

The structure of a resin particle is usually shown as in Figure 2, a network lined with linked but fully ionized sulphonic acid groups. The picture is a bit idealistic and somewhat misleading Figure 3 shows a much more satisfactory picture, as it emphasizes that not only sulphonic acid groups but also aliphatic chains and aromatic nuclei offer a surface to interact with species in solution.

Here now are a few early highlights of what ion exchange resins can do.

It has proved possible to prepare *solutions of unstable acids* without resorting to distillation or precipitation etc. A short column of the hydrogen

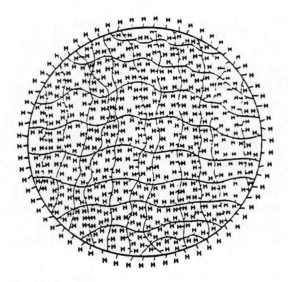

Figure 2. Basic structure of Dowex 50

form of a sulphonic resin will convert, for example, KCNS to HCNS, and the solution will not change concentration by this process.

Insoluble precipitates such as calcium oxalate can be decomposed at room temperature by allowing the solid to stand in water with an excess of the hydrogen form of a sulphonic resin overnight with occasional shaking. The Ca^{2+} ions will be held on the resin, and a solution of oxalic acid can be filtered or centrifuged off.

Sulphonic resins can be used as catalysts in esterification reactions (instead of sulphuric acid) and the resin is then readily separated by decantation.

All these early findings were exciting and firmly fixed the idea that a sulphonic resin is simply an insoluble form of sulphuric acid for most intents and purposes.

Techniques and equipment

Kurt A. Kraus, who did much of the pioneering work on the ion exchange of inorganic ions, maintained that you need no equipment. He did most of his work using 10 ml graduated pipettes as columns (the top constriction being sawn off and the tip plugged with glass wool). The eluent is added with a Pasteur pipette. The resin bed volume and the elution rate can be read from the graduations of the pipette. The eluate is then collected by drop counting on watch glasses; the fractions are evaporated on a water bath and their radioactivity counted with a mica end-window Geiger

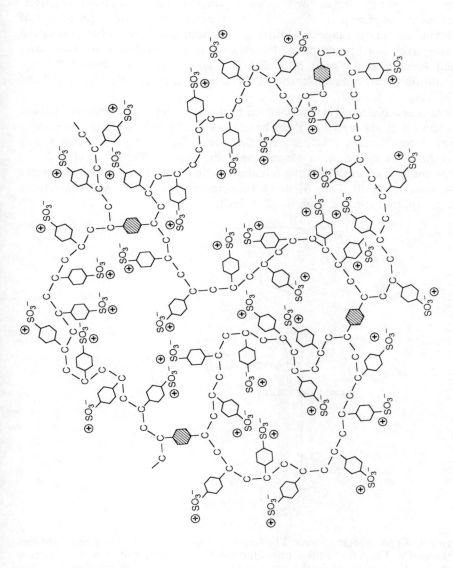

Figure 3. Structure of a polystyrene resin with sulphonic groups [7]. (Reproduced by permission of Ellis Horwood, Chichester)

counter. Kraus worked in Oak Ridge and most of his data were compiled using radioactive tracers which were readily available.

For large resin particles, columns as shown in Figures 4 and 5 were used in order to prevent the resin bed from running dry. In connection with rare earth separations, one of the first automatic detectors and also a 'degasser' were used, as shown in Figure 6. The type of arrangement consisting of column, automatic reagent addition and automatic photometric evaluation is extensively used in amino acid analysis and was developed by Moore and Stein (who got a Nobel prize for this work). It was adapted for the separation of condensed phosphates on anion exchange columns by Pollard et al. [11].

Kraus and Nelson [12] compiled the first chromatographic Periodic Tables. We shall only show two of these here, namely, those for anionic quaternary resin Dowex 1 and for sulphonic resin Dowex-50. See Figures 7 and 8. For an exhaustive collection of chromatographic Periodic Tables see [14]

On the *anionic resin*, alkalis, alkaline earths and Group 3A elements are not adsorbed at all. They do not form complexes with HCl and hence there are no anionic species, even in 12 N HCl.

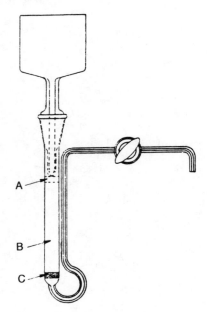

Figure 4. Experimental column. The funnel is removable to facilitate resin addition and removal. The resin bed B in this column is 1 cm^2 × 10 cm and rests on the porous glass disc C. A stopcock in the outlet tube allows regulation of the flow rate. The opening in the outlet tube is above the top of the resin bed, thus maintaining a liquid layer A above the resin at all times. (Reprinted with permission from Tompkins et al. [8]. Copyright (1947) American Chemical Society)

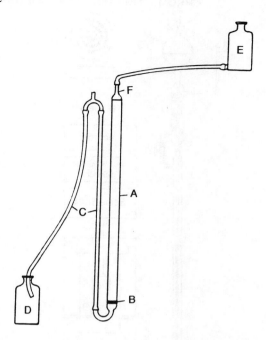

Figure 5. Apparatus after Harris and Tompkins [9]. The resin bed A rests on a porous disc B in the glass column F. The flow rate is adjusted by varying the height of the bottle E, which contains the influent solution. The effluent is collected in bottle D. The vent in tube C ensures a continuous liquid layer over the resin bed. (Reprinted with permission from Harris and Tompkins [9]. Copyright (1947) American Chemical Society)

Bromide and iodide show how a simple anionic monovalent ion is adsorbed. The K_d values are not high and the mass action effect desorbs them with increasing HCl concentration. ReO_4^- and TcO_4^- are more strongly adsorbed, as they are larger and hence less hydrated. For further discussion of TcO_4^- see pages 181–187. Zinc (II), Cd(II) and Hg(II), except at low concentrations of HCl, are present as MCl_4^{2-} complexes which, being poorly hydrated, are strongly retained and are desorbed by increasing the HCl concentration.

Then we have a number of metal ions which are complexed only at relatively high HCl concentrations: Fe(III), Co(II), Cu(II), Pa(V), UO_2^{2+} etc. We must emphasize that some anionic metal complexes have extremely high K_d values: Au(III), Ga(III) and Fe(III) above 6 N HCl, and Sb(V), which all form anionic chloro-complexes readily extracted into ether and strongly adsorbed on cellulose.

On the *cationic resin*, we note that halides are scarcely adsorbed; alkalis, alkaline earths and Group 3A metals are adsorbed but their adsorption is decreased readily by the mass action effect, i.e. with increasing HCl concentration up to about 5 M HCl, and then increases again. What happens at

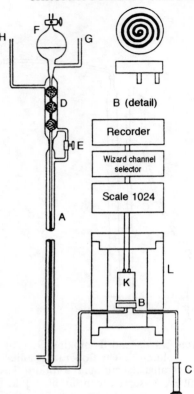

Figure 6. Experimental arrangement employed in ion exchange column separations with radioactive rare earths (Ketelle and Boyd [10]). A, adsorbent bed, Amberlite IR-1 or Dowex-50; B, counting cell; C, receiver, D, Allihn condenser; E, throttle valve; F, gas entrainment bulb; G, eluent inlet; H, thermostat fluid inlet; K, mica end-window Geiger–Müller counting tube; L, two-inch lead radiation shield. (Reprinted with permission from Ketelle and Boyd [10]. Copyright (1947) American Chemical Society)

high HCl concentration may be considered as follows. The amount of HCl is so high that the H_2O concentration is insufficient for complete hydration of H^+ and Cl^-. Thus there is a solution which is akin to an organic solvent and reduces the full hydration of the metal ions. The metal ions without full hydration are then bound more strongly to the sulphonic resin, either by electrostatic attraction or, more likely, by complex or ion pair formation with the sulphonic groups. The metals Au(III), Ga(III) and Fe(III) above 6 N HCl, and Sb(V), which are all strongly retained on the anionic resin, are also retained (although never as strongly) on the sulphonic resin. They all form anionic complexes or ion pairs of the type $[H^+AuCl_4^-]$ which certainly do not interact in any way with a–$SO_3^-H^+$ group. They are thus best considered as interacting with the organic network of the resin.

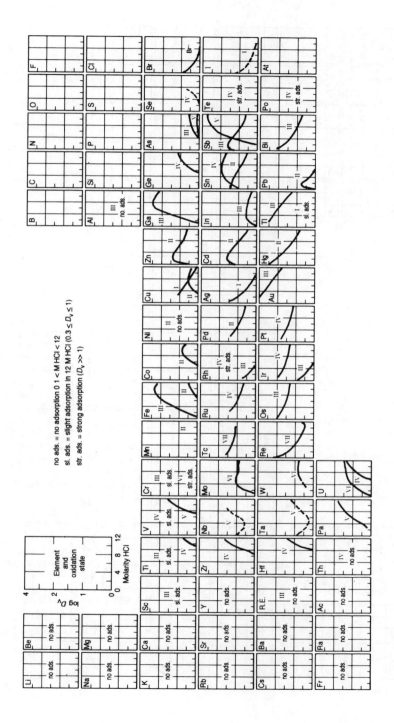

Figure 7. Anion exchange data for Dowex 1-X10 anion exchange resin with hydrochloric acid. (Kraus and Nelson [12])

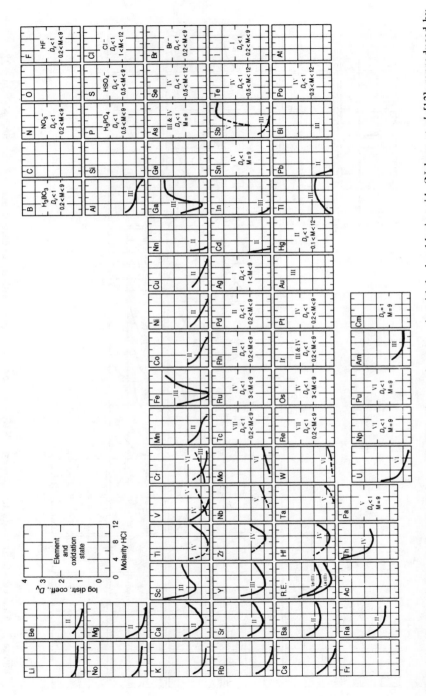

Figure 8. Cation exchange data for Dowex 50-X4 cation exchange resin with hydrochloric acid. (Nelson *et al.* [13]; reproduced by permission of Elsevier Science Publishers BV)

Below we give a selection of the many excellent separations which K. A. Kraus and co-workers have reported (Figures 9–11).

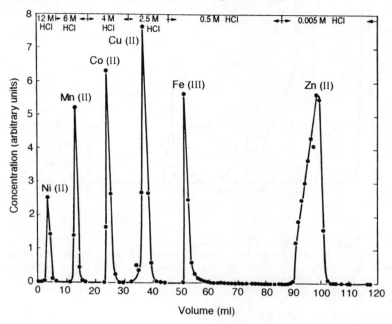

Figure 9. Separation of transition elements Mn to Zn. Dowex 1 column, 26 cm × 0.29 cm^2; flow rate 0.5 cm/min. (Reprinted with permission from Kraus and Moore [15]. Copyright (1953) American Chemical Society)

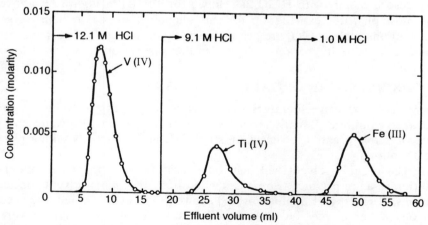

Figure 10. Separation of V(IV), Ti(IV) and Fe(III) by anion exchange. Dowex 1 column, 18.6 cm × 0.49 cm^2; 1 ml = 0.05 M V(IV), 0.018 M Ti(IV), 0.025 M Fe(III) in 12 M HCl. (Reprinted with permission from Kraus and Moore [16]. Copyright (1950) American Chemical Society)

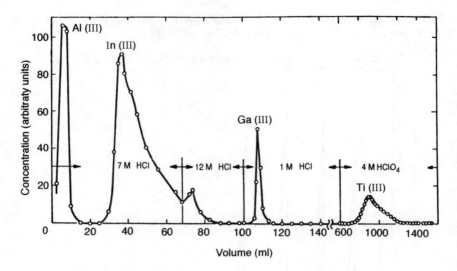

Figure 11. Separation of Al(III), Ga(III), In(III) and Tl(III). Dowex 1 column, 20 cm × 0.4cm^2, room temperature, flow rate 0.3–0.8 cm/min (Kraus et al. [17]; reprinted with permission from *J. Phys. Chem.* **58** (1954) 11; copyright (1954) American Chemical Society)

THE PERCHLORATE EFFECT

Some of the species very strongly adsorbed on anion exchangers would require very large volumes of HCl for elution. It was found, however, that they desorb readily with $HClO_4$, as shown for Tl(III) in Figure 11, above. The most plausible explanation for this phenomenon is that high oxygenated anions such as ClO_4^- are just as poorly hydrated as Tl(III)Cl_4^- and can compete successfully for ion exchange and/or adsorption sites.

ION EXCHANGE IN ORGANIC SOLVENTS

There is an extensive literature on the effect of organic solvents on ion exchange equilibria. Usually mixtures of an acid or a salt with water and a water-miscible organic solvent were studied. The main events taking place are as follows.

(i) The organic solvent will distribute itself between the resin and the solution so as to produce two phases with different organic solvent concentrations. Rückert and Samuelson [18] showed that, at low ethanol concentrations, ethanol may be preferentially adsorbed on the resin network, whereas at high ethanol concentrations water will be preferentially adsorbed by the ionized exchange groups. See Figure 12. Thus a solvent–solvent extraction equilibrium is superimposed on the usual ion exchange and adsorption equilibria.

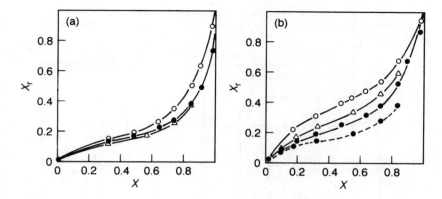

Figure 12 (a). Distribution of ethanol between various forms of Dowex 50 X-8 and water. Mole fraction of ethanol (X_r) in the resin phase as a function of mole fraction in solution (X). ○, Li; △, Na; ●, K [18] (b). Distribution of ethanol between various forms of Dowex 2-X8 (——) and Dowex 2 X-1 (----) and water. Mole fraction of ethanol (X_r) in the resin phase as a function of mole fraction in solution (X). ○, ClO_4^-; △, Cl^-; ●, SO_4^{2-} [18]

(ii) Hydrated ions such as Li^+ or Na^+ are less strongly hydrated in the presence of organic solvents and will be more readily adsorbed on cation exchangers.

(iii) Reversible complexing equilibria such

$$Cd_{aq}^{2+} \rightleftharpoons CdCl^+ \rightleftharpoons CdCl_2^0 \rightleftharpoons CdCl_3^- \rightleftharpoons CdCl_4^{2-}$$

will be shifted towards the complexed form as the concentration of H_2O is diminished. Thus metals will be adsorbed on anion exchangers at lower complexant concentration.

(iv) Very strongly adsorbed species such as $[H^+AuCl_4^-]$ will be desorbed from a sulphonic resin in the presence of organic solvents.

A splendid example of how organic solvents can be used for a multi-metal separation is shown in Figure 13.

ION EXCHANGE CELLULOSES

Most of the fundamental work on ion exchange of inorganic species has been carried out with resins having a styrene–divinylbenzene network. Biochemists soon found that this type of resin was also fine for small molecules such as amino acids, nucleobases or sugars (as borate complexes) but generally failed with proteins, which were too strongly adsorbed and often denatured.

Carboxylic exchangers on a cellulose network proved successful for protein separations and soon also cellulose with sulphonic groups, DEAE-cellulose (diethylaminoethylcellulose) and AE-cellulose (aminoethylcellulose) became

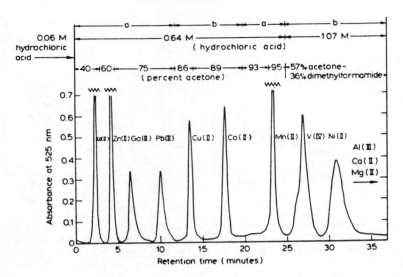

Figure 13. Separation of metal ions on Amberlite 200 (25–30 µm). Column, 120 mm × 5 mm i.d.; flow rate, 22 cm/min; temperature, 40 °C; column inlet pressure, 16–25 atm; colour-forming reagent solution, 0.02% 4-(2-pyridylazo) resorcinol in ammonia, concentration (a) 0.7 M and (b) 1.2 M; sample volume, 60 ml. Amount of metal (× 10^{-8} mol): Cd^{2+}, 9.6; Zn^{2+}, 4.8; Ga^{3+}, 4.5; Pb^{2+}, 12; Cu^{2+}, 4.8; Co^{2+}, 7.2; Mn^{2+}, 12; V^{4+}, 57; Ni^{2+}, 12; Al^{3+}, 9.6; Ca^{2+}, 1.2 × 10^3; Mg^{2+}, 6 × 10^2 [19]. (Reproduced by permission of Elsevier Science Publishers BV)

commercially available as powders for columns, as papers and later also as thin layers.

These media are much less stable than resins, and columns cannot be put under pressure. Figure 14 illustrates the differences between resin papers, DEAE-cellulose papers, AE-cellulose papers and cellulose paper for many elements of the Periodic Table.

Figure 14. Anion exchange paper chromatography. R_F values of elements in aqueous hydrochloric acid on three anion exchange papers: SB-2 resin paper, DE-20 cellulose anion exchange paper and AE-30 anion exchange paper, and on Whatman No. 1 paper (for comparison) [20]. Top, left: the theoretical R_F–H^+ curve of a monovalent anion obeying the law of mass action (calculated for three different R_F values at 1 N H^+). Top, right: an alignment chart for converting R_F values on SB-2 paper to R_F values on AE-30 or DE-20 paper (or vice versa). In this alignment chart only the differences in water/paper ratios and exchange capacities were considered. Below: R_F values of metal ions on Whatman No. 1 paper (○——○), SB-2 resin paper (■——■), DE-20 cellulose anion exchange paper (●——●) and AE-30 cellulose anion exchange paper (□——□) are plotted against the concentration of HCl. With the exception of Pt(II), which is placed below Pt(IV), the metal ions are placed in their positions in the Periodic Table. (Reproduced by permission of Elsevier Science Publishers BV)

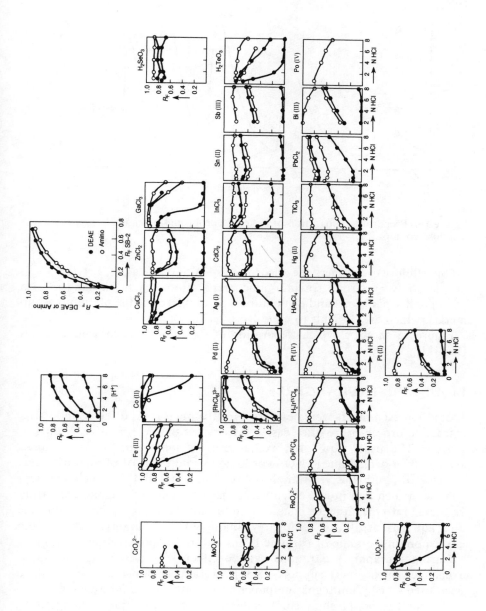

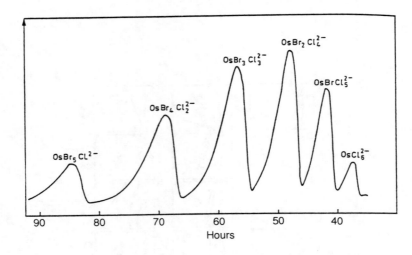

Figure 15. Preparative separation of mixed chlorobromo-osmiates(IV) on a DEAE-cellulose column [21]

On cellulose exchangers the adsorption of Au(III), Ga(III), Fe(III), Sb(V) etc. is not very strong. In inorganic preparations, these exchangers also yield good separations and fast elution for complexes of the platinum group metals, which are difficult to elute from resin columns.

Figure 15 shows a typical separation of the mixed chlorobromoosmates (IV) [21].

LIQUID ION EXCHANGERS

Liquid ion exchangers were developed for the nuclear industry, as industrial extraction processes are often easier to carry out than preparative chromatography. The usual liquid ion exchangers are long-chain tertiary amines such as tri-n-octylamine (TNOA) or long-chain diesters of phosphoric acids such as di-(2-ethylhexyl)orthophosphoric acid (HDEHP), which are very soluble in such non-polar solvents as benzene or chloroform and are hardly extracted into the aqueous phase. Equilibria between such organic solutions and aqueous acids are completely analogous to equilibria between the corresponding anionic or cationic resins and the same aqueous solution.

Chromatographic separations can be effected by adsorbing the liquid exchanger on an inert support—Kel-F, Teflon, cellulose and so on have been used—or by impregnating paper strips with the liquid exchanger.

Figures 16 and 17 show some example where separations analogous to those on resin columns were obtained. A good survey of the extensive literature on this topic is found in the book by Braun and Ghersini [22].

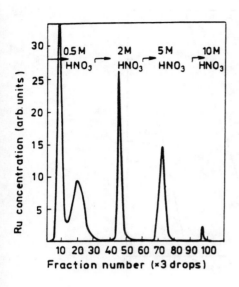

Figure 16. Separation of RuNO-nitrato complexes aged in 3 M HNO$_3$. TB- kieselguhr column, 120 mm × 3.5 mm diam. Temperature 0 °C; flow rate 8–10 drops/min; fraction volume 90 µl [22]. (Reproduced by permission of Elsevier Science Publishers BV)

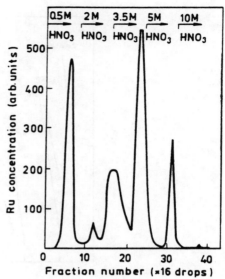

Figure 17. Separation of RuNO-nitro complexes aged in 5 M KNO$_3$. TBP-kieselguhr column, 120 mm × 3.5 mm diam. Temperature 0 °C; flow rate 10–12 drops/min; fraction volume 0.32 ml [22] (Reproduced by permission of Elsevier Science Publishers BV)

INORGANIC EXCHANGERS

Silica

Silica, silica gel and silicic acid are synonyms for insoluble SiO$_2$, which is the most widely used adsorbent in the chromatography of organic compounds. On its surface there are mainly silanol groups Si=OH and (if heated above 110°C) siloxane groups SiO as well as some diol groups $Si{<}^{OH}_{OH}$.

In the absence of water, silanol groups behave as exchangers of hydrogen bonds and siloxane groups are weakly hydrophobic. In the presence of water, silanol groups ionize as weak acids and can function as cation exchangers as well as forming ion pairs with metal ions. Distribution coefficients have been measured as shown in Figure 18.

Above pH 8 silica dissolves readily on the surface, forming silicate anions.

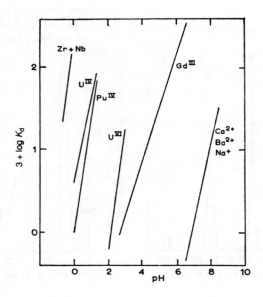

Figure 18. Distribution coefficients for various ions on silica gel as a function of pH. SiO_2: Kebo 50–100 mesh. Zr, Nb 1.1 mM; U^{IV} 2.0; Pu^{IV} 0.01; UVI 1.0; Gd^{III} 0.092; Ca^{2+} 25; Ba^{2+} 25; Na^+ 30. (From [23], p. 90; reproduced by permission of Elsevier Applied Science Publishers Ltd.)

Alumina

Al_2O_3 will ionize in acid solution to yield on the surface $>Al^+$ and in alkaline solutions to yield $>Al–O^-$. We have already mentioned that Schwab and co-workers (see page 8) obtained sequences of metal ions and of anions on alumina.

The mechanism of the interaction of cobaltammines with alumina was discussed by Sacconi [24], and he suggested a hydrolytic process on the surface of alumina. Bjerrum and his co-workers [25] adopted chromatography on alumina as a general method for separating various species in reactions of Co(III) complexes. They argued that they preferred a white adsorbent to the strongly coloured resins.

Zirconium oxide

Inorganic exchangers were developed for the nuclear industry when it was realised that organic resins decomposed when highly radioactive solutions were chromatographed.

Zirconium oxide was proposed by K. A. Kraus et al. [26] as a highly insoluble metal oxide, and it was shown that the equilibrium nitrate–bromide (or nitrate–chloride) follows the law of mass action. Chromate was found to be adsorbed very strongly.

Sakodynskii and Lederer [27] noted that some other ions are not desorbed with increasing salt concentration as would be expected by the law of mass action; there thus seem to be further mechanisms of adsorption.

Titanium hydroxide or titanium dioxide

This compound, precipitated on muslin or glass wool, was recommended by Davies et al. [28] for the removal of uranium from sea water. It was also investigated by Sakodynskii and Lederer [27] and, again, the adsorption data cannot be explained by ion exchange alone.

Actually, all insoluble hydroxides and oxides of metals, e.g. SnO_2, ThO_2 and $Fe(OH)_3$, will exhibit ion exchange and other adsorption on their surface (Figure 19). This phenomenon is usually not sufficiently stressed in the teaching of analytical methods, that a precipitate of $Fe(OH)_3$ for example will strongly interact with all metal ions present.

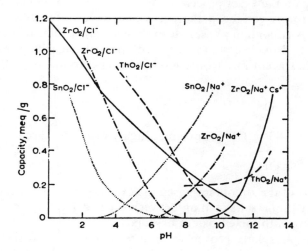

Figure 19. Anion and cation exchange capacities of hydrous ZrO_2, ThO_2 and SnO_2 as a function of pH. (From [23], p. 90; reproduced by permission of Elsevier Applied Science Publishers Ltd.)

For further reading, the small book by Amphlett [23] can be recommended.

Zirconium phosphate and other insoluble salts of trivalent and tetravalent metals

Zirconium phosphate was first given the formula shown in Figure 20, which was later questioned on the basis of analyses which showed that the formula $Zr(HPO_4)_2 \cdot n\ H_2O$ is more likely, i.e. that the Zr is bonded by OH bridges to a hydrated Zr rather than by Zr–O–Zr. It can be prepared by precipitation from a solution of zirconyl chloride with phosphoric acid.

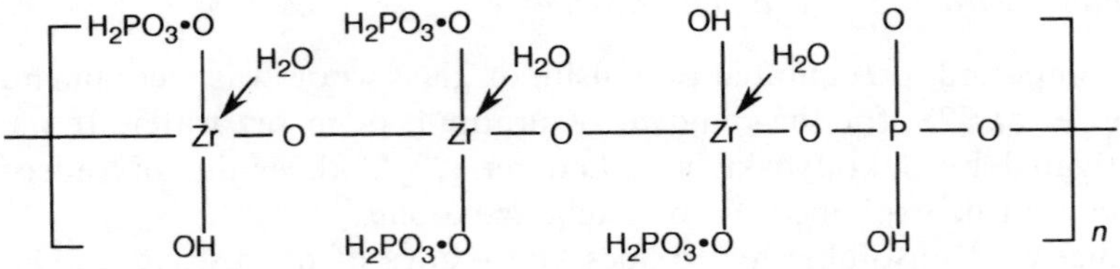

$$\left[\begin{array}{c} H_2PO_3\cdot O \\ | \\ Zr-O \\ | \\ OH \end{array} \xrightarrow{H_2O} \begin{array}{c} H_2PO_3\cdot O \\ | \\ Zr-O \\ | \\ H_2PO_3\cdot O \end{array} \xrightarrow{H_2O} \begin{array}{c} OH \\ | \\ Zr-O \\ | \\ H_2PO_3\cdot O \end{array} \xrightarrow{H_2O} \begin{array}{c} O \\ \| \\ P-O \\ | \\ OH \end{array} \right]_n$$

Figure 20. A proposed formula for zirconium phosphate. (Reproduced by permission of Elsevier Applied Science Publishers Ltd.)

Depending on the exact procedure of its preparation, either crystalline or amorphous forms can be prepared, which have different capacities and selectivities. A zirconium phosphate exchanger can also be made by washing a zirconium hydroxide column with a solution of phosphate. Although the products differ according to the mode of preparation, they all possess on the surface

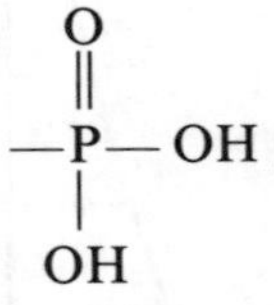

$$\begin{array}{c} O \\ \| \\ -P-OH \\ | \\ OH \end{array}$$ groups in a highly insoluble form.

In accounts of the zirconium-based ion exchangers, it is often ignored that the usual zirconium salts contain between 0.1 and 7% of hafnium, and thus we are dealing with mixed oxides or salts.

Zirconium phosphate proved to be a very interesting cation exchanger. Some R_F values for zirconium phosphate-impregnated papers are given in Table 1. There is good selectivity for the alkalis, and this has been used in nuclear chemistry to separate Cs^+ from many other metal ions.

The law of mass action holds for most exchange reactions so far examined; see Figure 21.

Most salts of trivalent and tetravalent metal ions form precipitates with polyvalent anions and yield ion exchangers.

ION EXCHANGE   105

Table 1. R_F values of inorganic cations on zirconium phosphate paper [29]. (Reproduced by permission of Elsevier Science Publishers BV)

| Elements[a] | \multicolumn{2}{c}{HCl} | \multicolumn{2}{c}{HNO$_3$} | \multicolumn{2}{c}{H$_2$SO$_4$} | \multicolumn{2}{c}{HClO$_4$} |

Elements[a]	HCl 0.1 N	HCl 1 N	HNO$_3$ 0.1 N	HNO$_3$ 1 N	H$_2$SO$_4$ 0.1 N	H$_2$SO$_4$ 1 N	HClO$_4$ 0.1 N	HClO$_4$ 1 N
Li(I)	0.76	0.85	0.77	0.87	0.77	0.84	0.79	0.80
Na(I)[b]	0.67	0.83	0.75	0.84	0.50	0.77	0.66	0.78
K	0.50	0.60	0.40	0.75	0.30	0.64	0.66	0.78
Rb(I)	0.12	0.48	0.15	0.50	0.07	0.34	0.13	0.47
Cs(I)[b]	0.00	0.00	0.00	0.00	0.00	0.00	0.00	0.00
Ag(I)	–	–	0.00	0.05	0.00	0.00	–	0.00
Tl(I)	0.00	0.45	0.05	0.30	0.10	0.25	0.08	0.30
Ca(II)[b]	0.80	–	0.66	0.81	–	–	0.71	0.82
Sr(II)[b]	0.77	–	0.62	0.84	–	–	0.68	0.80
Co(II)	0.67	0.77	0.62	0.80	0.53	0.87	0.50	0.73
Zn(II)	0.63	0.76	0.47	0.77	0.64	0.80	0.45	0.78
Ni(II)	0.61	0.85	0.64	0.86	0.63	0.86	0.53	0.80
Cd(II)	0.60	0.81	0.40	0.79	0.46	0.86	0.34	0.80
Cu(II)	0.56	0.78	0.48	0.81	0.41	0.84	0.46	0.83
Pb(II)	–	–	0.08	0.75	–	–	0.08	0.53
Al(III)	0.13	0.77	0.00	0.75	0.02	0.72	0.05	0.71
Fe(III)	0.00	0.04	0.00	0.00	0.00	0.02	–	0.00
Bi(III)	–	0.83	0.06	0.28	0.00	0.00	–	0.00
La(III)	0.28	0.84	0.28	0.80	0.20	0.82	0.20	0.80
Ce(III)	0.25	0.84	0.15	0.76	0.10	0.71	0.05	0.78
Y(III)	0.05	0.65	0.06	0.66	0.01	0.67	0.00	0.58
Th(IV)	0.00	0.00	0.00	0.00	0.04	0.04	0.00	0.00
UO$_2$(II)	0.00	0.32	0.00	0.15	0.12	0.29	0.02	0.20

[a] The spots were detected by spraying the strips with suitable reagents.
[b] These elements were also detected by radiochemical methods.

Numerous such exchangers have been described but, unless they are formed under well controlled conditions, they are irreproducible with respect to both selectivity and ion exchange capacity.

Figures 22–24 show some separations which have been achieved on zirconium tungstate and zirconium molybdate.

Insoluble inorganic acids (heteropolyacids)

The most interesting compounds of this group are phosphomolybdic acid and polyantimonic acid. Both are highly insoluble and show interesting selectivities in chromatography.

A separation of the alkalis on paper impregnated with ammonium phosphomolybdate (AMP) is shown in Figure 25, where Li and Na are separated by an additional development with 95% ethanol. A column separation is shown in Figure 26. A separation of alkalis on crystalline

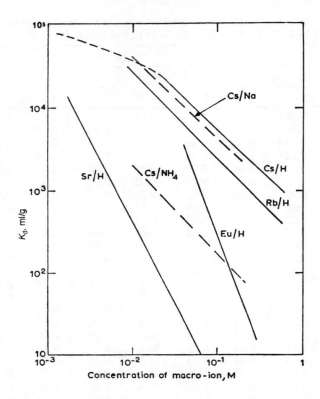

Figure 21. Distribution coefficients for tracer ions on zirconium phosphate in the presence of different macro-ions. (From [23], p. 105; reproduced by permission of Elsevier Applied Science Publishers Ltd.)

antimonic acid is shown in Figure 27. Here the affinity for Na^+ is quite remarkable.

Inorgano-organic ion exchangers

Active research is proceeding on mixed zirconium phosphate–organic acid precipitates. Benzene phosphonate and other organic phosphates with various substituents have been included in zirconium phosphate. These compounds show a wide spectrum of ion exchange and intercalation behaviour, e.g. Figure 28. For a review see [35].

ION EXCHANGE PAPERS AND THIN LAYERS

We have mentioned ion exchange papers already in several sections of this chapter, namely those with functional groups covalently linked to the

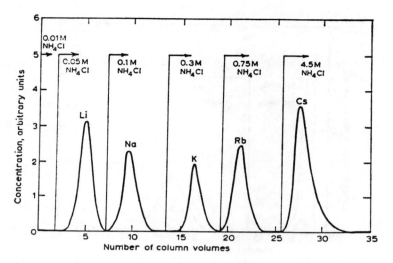

Figure 22. Separation of the alkali metals at tracer concentration on zirconium tungstate. Column 12.3 cm × 0.13 cm^2; flow rate ≈0.75 cm/min [30]

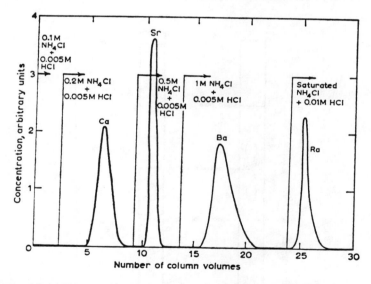

Figure 23. Separation of the alkaline earths at tracer concentrations on zirconium molybdate. Column 10.0 cm × 0.19 cm^2; flow rate 1.1 cm/min [31]

cellulose network. Whatman markets a cellulose phosphate paper, a carboxymethylcellulose paper, an aminoethylcellulose paper, a dimethylaminoethylcellulose paper and an Ecteola-cellulose paper. They have all rather low capacities and are not stable indefinitely. They have yielded excellent

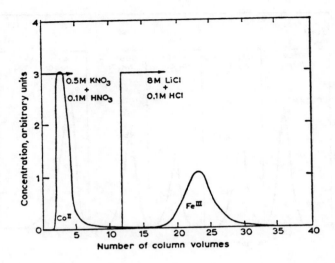

Figure 24. Separation of cobaltous and ferric ions on zirconium tungstate. Column 5.5 cm × 0.09 cm^2; flow rate 0.5 cm/min [31]

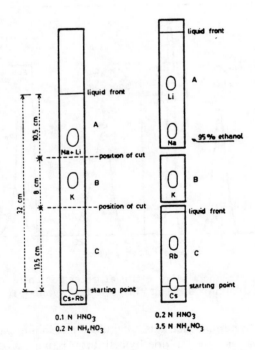

Figure 25. Separation of alkali metal ions on AMP-impregnated paper [32]. (Reproduced by permission of Elsevier Science Publishers BV)

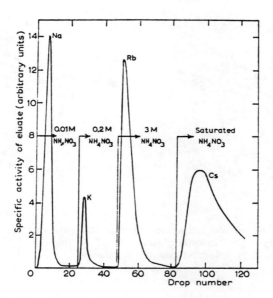

Figure 26. Separation of alkali metals at tracer concentrations on a bed of ammonium molybdophosphate; diameter 5 mm, depth 1.6 mm [33]

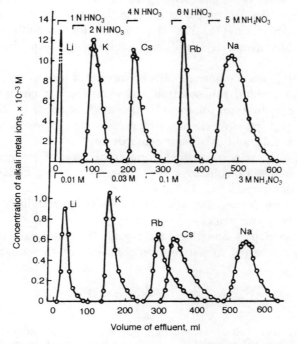

Figure 27. Separation of alkali metal ions on crystalline antimonic acid with nitric acid and ammonium nitrate as eluents [34]

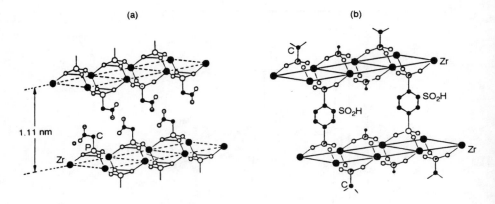

Figure 28. (a) Schematic structure of an inorgano-organic ion exchanger, $Zr(PO_3CH_2COOH)_2$. (b) Schematic structure of a 'pillared' ion exchanger with sulphonic groups [35]. (Reproduced by permission of Elsevier Applied Science Publishers Ltd.)

chromatograms, especially of transition metal complexes which are adsorbed too strongly on resins (see Figure 14).

Resin papers can be prepared by dipping papers into a suspension of finely divided resin [36, 37]; also, papers loaded with 45% of resin are commercially available (sulphonic, carboxylic, strong base and weak base).

Papers holding metal oxides or salts are prepared for example, by dipping, the paper into a metal salt solution and exposing the paper to ammonia fumes (for the hydroxide) followed by thorough washing or dipping into a suitable acid (e.g. phosphoric acid) to precipitate the insoluble salt inside the paper (also followed by washing and drying).

Work with such papers became popular because of its speed and simplicity compared with column work. Also, radioactive zones are readily detected on such papers (see Table 1).

Analogous thin layers are also available commercially or have been 'home-made' by some workers.

Ion exchange papers and thin layers yield the same data as columns if R_F values are converted to distribution constants by the equation

$$\left(\frac{1}{R_F} - 1\right)\frac{A_L}{A_S} = \alpha = D$$

Some comparative values are shown in Table 2. The accuracy of the D values obtained from R_F data is considerably lower than that of D values from equilibrium experiments.

Table 2. Comparison of R_F values measured on SB-2 resin paper and calculated from equilibrium data from Kraus *et al.* on Dowex 1 resin

Concentration of HCl	R_F values of Cu(II)			R_F values of Fe(III)		
	measured	calculated from D	D	measured	calculated from D	D
2 N...	0.49	–	1.92	0.71	0.735	0.6
5 N...	0.03	0.027	67	0.29	0.39	2.9
6 N...	0.04	0.004	460	0.21	0.17	9.6
8 N...	0.0	–	–	0.16	0.17	9.6
9 N...	–	0.0003	5900	–	–	–

ESTIMATION OF THE 'CHARGE' ON AN ION BY ION EXCHANGE DATA

In early work on ion exchange resins, it was believed that adsorption on a cation exchanger was conclusive evidence that a complex or species was cationic. Thus, Ayres [38] found that platinum metal–stannous chloride complexes were adsorbed on a cation exchanger and based a cationic 'formula' on this observation. If we look at Figures 7 and 14 we see that a number of anionic complexes are adsorbed on Dowex-50. Shukla [39] showed subsequently that the platinum metal–stannous chloride complexes migrated as anions in paper electrophoresis, and this led later to their identification as anions.

Similarly, it was believed that the charge on a complex could be estimated from the law of mass action equation of the adsorption of an ion on a resin. The best summary of this work can be found in a review by Carunchio and Grassini [40]:

Total charge on the complex
To determine this quantity, most authors have used methods very similar to or the same as that developed by Nelson and Kraus [41] for investigating alkaline earth metal citrates on anion exchangers. In this method, the ion exchange equilibrium is expressed as:

$$aX^{x-} + x(A^{a-})_r \rightleftharpoons a(X^{x-})_r + xA^{a-} \qquad (4)$$

where X^{x-} and A^{a-} denote respectively the complex ion and the eluting anion, the resin phase being indicated by r. The exchange constant can then be written as:

$$K = G \frac{[X^{x-}]_r^a [A^{a-}]^x}{[X^{x-}]^a [A^{a-}]_r^x} \qquad (5)$$

where G is the ratio of the activity coefficients and the quantities in brackets are concentrations. Assuming that G and the term $[A^{a-}]_r$ are constant, i.e. that $[X^{x-}]_r$ is small in comparison with $[A^{a-}]_r$, we can write equation in the form:

$$\text{const.} = \frac{K}{G}[A^{a-}]_r^x = D^a[A^{a-}]^x \tag{6}$$

since:

$$D = \frac{[X^{x-}]_r}{[X^{x-}]} \tag{7}$$

Differentiating equation (6), we obtain:

$$\frac{d \log D}{d \log [A^{a-}]} = S = -\frac{x}{a} \tag{8}$$

The slope S of the curve obtained by plotting log D against log $[A^{a-}]$ is numerically equal to the ratio between the charge on the complex ion (x) and that on the eluting anion (a).

Li and White used this method to determine the charge on the citrates of Co(II), Zn(II) and Th(IV) and verified the result for the Co(II) citrate by a somewhat different technique, in which the resin was not in the citrate but in the Cl⁻ form. However, the attempts to determine x for a uranyl citrate complex failed, since the very strong affinity of the uranyl ion for the resin citrate makes it impossible to obtain reasonable distribution coefficients for this system. The very stable uranyl citrate complex, which is evidently formed, may serve as a basis for the separation of uranium from metals on anion exchangers.

The method of Nelson and Kraus has been used in conjunction with other techniques to determine the charge x on several complexes such as the oxygenated cobalt glycylglycinate ($x = -1$), a plutonium (IV) oxalate ($x = -6$), two protactinium (V) complexes formed in H_2SO_4 solutions ($x = -1$ and -3), complexes of Sr(II) with dicarboxylic acids ($x = 0$), and a Pm(III)-EDTA complex formed in solutions with pH 1.8–9.0 ($x = -1$). The value of x has also been determined for some complexes of Fe(II) and Ti(IV) by Beukenkamp and Herrington using a similar method and by Nabivanets for the complexes of Ti(IV) formed in HCl solutions of various concentrations.

A cation exchanger in the H⁺ form is used in the method of Cady and Connick for determining x of cationic complexes. This method is based on the fact that the equilibrium distribution of the complex cation depends on the concentration of H⁺ (the exchangeable ion) in the aqueous phase and the resin phase. Thus the ion exchange constant K for the reaction:

$$M^{x+} + x(H^+)_r \rightleftharpoons (M^{x+})_r + xH^+ \tag{9}$$

is given by:

$$K = G \frac{[M^{x+}]_r [H^+]^x}{[M^{x+}][H^+]_r^x} \tag{10}$$

where G is the ratio of the activity coefficients. The distribution of M^{x+}, i.e. $[M^{x+}]_r/[M^{x+}]$, varies as the x-th power of the hydrogen ion distribution $[H^+]_r/[H^+]$. We can calculate x and K by substituting in (10) the values obtained from two equilibrations at different hydrogen ion concentrations. Strictly accurate values are obtained only if K is the same for both equilibrations. This is not so in practice, because the activity coefficients in the aqueous phase vary from one equilibration to another. However, the method can still be used, since x is known to be a whole number of n (charge per metal atom), so the experimental values need not be very accurate.

Lederer and Kertes correlated the charge on reversibly adsorbed cationic complexes with their R_F values on ion exchange paper. Equations (9) and (10) are equally valid for exchange reactions occurring on paper impregnated with ion exchangers. The logarithmic form of equation (10) is as follows (provided that $G = 1$):

$$\log K = \log \frac{[M^{x+}]_r}{[M^{x+}]} + x \log [H^+] - x \log [H^+]_r \quad (11)$$

To express $\log K$ as a function of R_F, the only parameter that is directly measurable on paper, we use the following relationship valid for partition chromatography:

$$R_F = \frac{A_L}{A_L + \alpha A_S} \quad (12)$$

where A_L is the cross-section of the mobile phase, A_S is the cross-section of the stationary phase, and α is the partition coefficient (ratio of the concentration in the stationary phase to that in the mobile phase). Equation (12) can also be written as:

$$\frac{1}{R_F} - 1 = \alpha \frac{A_S}{A_L} \quad (13)$$

Substitution of this into equation (II) gives:

$$\log K = \log \frac{A_L}{A_S} + \log \left[\frac{1}{R_F} - 1 \right] + x \log [H^+] - x \log [H^+]_r \quad (14)$$

Moreover, since $\log K$, A_L/A_S, and $x \log [H^+]_r$ are assumed to be constant, we obtain:

$$\log \left[\frac{1}{R_F} - 1 \right] = - x \log [H^+] + \text{constant} \quad (15)$$

Utilizing the relationship $(1/R_F - 1) = R_M$ we may simplify equation (15) and write:

$$R_M = x \, \text{pH} + \text{constant} \quad (16)$$

This relationship holds for each given cation eluted with hydrogen ions but the method is limited to eluent concentrations that give rise to intermediate R_F values. It has been used with Al(III) in mixtures of HCl and HF of various concentrations. Moreover, equation (15) has been found valid when a paper impregnated with zirconium phosphate (K^+ form) is used and the elution is carried out with highly concentrated solutions of KCl. In this case, the concentration of H^+ is naturally replaced by that of K^+ in equation (15). [Note: equation numbers are part of the quotation]

However, the kind of argument presented above can not be generally applied [41]. Firstly, complexes like $Co(NH_3)_6^{3+}$ and $Co(en)_3^{3+}$ behave as if they had a charge of 5. It has been postulated that these complexes, which are known to form strong outer sphere complexes with sulphate, also form outer sphere complexes with the sulphonic groups of the resin [42].

It was shown furthermore, that the 'charge' can vary from 1.5 to 3 for trivalent rare earth metals depending on the 'concentration' of exchange groups on the surface of the exchanger.

There is also the question whether trivalent cations should exhibit a charge of 3+ in such law of mass action equilibria. Both conductivity and electrophoretic data seem to indicate that in 0.1 M solutions or higher trivalent cations are surrounded by a cloud of anions either by outer sphere complexing or simply by electrostatic effects, so that they cannot possibly be involved in an electrostatic equilibrium exhibiting a charge of 3+. Thus, for apparent equilibria showing a charge of more than 2+, the mechanism may be partly ion pairing and not simply electrostatic attraction.

SOME EXAMPLES OF APPLICATIONS

Separations of coordination complexes by simple ion exchange columns on a small preparative scale (about ten to some hundreds of milligrams) is now routinely used by coordination chemists. Such separations are usually not even reported separately but included in the paper reporting the investigation on the compounds.

Below we show several examples from the early literature. Stepwise elution with different eluents is often practised so as to achieve a separation without unduly long elution times. See Figures 29–31.

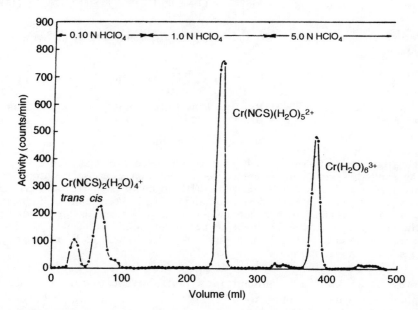

Figure 29. Elution curve of a mixture of Cr(III)–thiocyanate complexes from Dowex 50 cation exchange resin. (Reprinted with permission from Kaufman [43]. Copyright (1960) American Chemical Society)

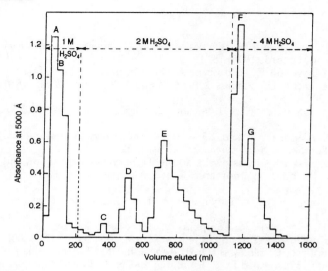

Figure 30. Ion exchange separation of cobalt species from the reaction of $[(NH_3)_5 CoOOCo(NH_3)_5](SO_4)_2$ with aqueous H_2SO_4. $B = Co^{2+}$; $F = [(NH_3)_5Co(H_2O)]^{3+}$; $G = [Co(NH_3)_6]^{3+}$; $H = [(NH_3)_5CoOOCo(NH_3)_5]^{5+}$ (fraction H was not eluted); A, C, D, E = unidentified fractions. (From Charles and Barnartt [44])

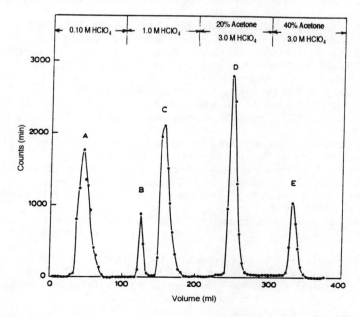

Figure 31. Elution curve of a mixture of Cr(III)–thiocyanate complexes from Selectacel DEAE anion exchange resin. $A = [Cr(NCS)_3(H_2O)_3]$; $B = trans$-$[Cr(NCS)_4(H_2O)_2]^-$; $C = cis$-$[Cr(NCS)_4(H_2O)_2]^-$; $D = [Cr(NCS)_5(H_2O)]^{2-}$ [45]. (Reprinted) with permission from *Anal. Chem.* **36** (1964) 1777. Copyright (1964) American Chemical Society)

References

[1] W. Rieman III and H. F. Walton, *Ion Exchange in Analytical Chemistry*, Pergamon, Oxford (1970), p. 1.
[2] B. A. Adams and E. L. Holmes, *J. Soc. Chem. Ind.* **54** (1935) 1.
[3] O. Liebknecht, US Patents 2,191,060 and 2,206,007; *Chem. Abstr.* **34** (1940) 4501, 7503.
[4] P. Smit, US Patents 2,191,063 and 2,205,635; *Chem. Abstr.* **34** (1940) 4500, 7504.
[5] G. F. D'Alelio, US Patent 2,366,007; *Chem. Abstr.* **39** (1945) 4418.
[6] E. R. Tompkins, *Anal. Chem.* **22** (1950) 1352.
[7] J. Minczewski, J. Chwastowska and R. Dybczynski, Separation and Preconcentration Methods in Inorganic Trace Analysis, Ellis Horwood, Chichester (1982), p. 285.
[8] E. R. Tompkins, J. X. Khym and W. E. Cohn, *J. Am. Chem. Soc.* **69** (1947) 2769.
[9] D. H. Harris and E. R. Tompkins, *J. Am. Chem. Soc.* **69** (1947) 2792.
[10] B. H. Ketelle and G. E. Boyd, *J. Am. Chem. Soc.* **69** (1947) 2800.
[11] F. H. Pollard, G. Nickless, D. E. Rogers and M. T. Rothwell, *J. Chromatogr.* **17** (1965) 157; see also D. P. Lundgren and N. P. Loeb, *Anal. Chem.* **33** (1961) 366.
[12] K. A. Kraus and F. Nelson, *Proc. First Int. Conf. Peaceful Uses At. Energy*, Geneva (1955), Vol. 7, pp. 113–125.
[13] F. Nelson, T. Murase and K. A. Kraus, *J. Chromatogr.* **13** (1964) 503.
[14] M. Lederer, *The Periodic Table for Chromatographers*, John Wiley, Chichester (1992).
[15] K. A. Kraus and G. E. Moore, *J. Am. Chem. Soc.* **75** (1953) 1460.
[16] K. A. Kraus and G. E. Moore, *J. Am. Chem. Soc.* **72** (1950) 4293.
[17] K. A. Kraus, F. Nelson and G. W. Smith, *J. Phys. Chem.* **58** (1954) 11.
[18] H. Rückert and O. Samuelson, *Acta Chem. Scand.* **11** (1957) 303.
[19] K. Kawazu, *J. Chromatogr.* **137** (1977) 381.
[20] M. Lederer and L. Ossicini, *J. Chromatogr.* **13** (1964) 188.
[21] H. Müller, P. Bekk and I. Hagenlocher, *Z. Anorg. Allg. Chem.* **503** (1983) 15.
[22] T. Braun and G. Ghersini, Extraction Chromatography, Elsevier, Amsterdam (1975).
[23] C. B. Amphlett, *Inorganic Ion Exchangers* Elsevier, Amsterdam, (1964).
[24] L. Sacconi, *Discuss. Faraday Soc.* **7** (1949) 173.
[25] A. Jensen, J. Bjerrum and F. Woldbye, *Acta Chem. Scand.* **12** (1958) 1047.
[26] K. A. Kraus, H. O. Phillips, T. A. Carlsen and J. S. Johnson, *Prog. Nuclear Energy*, Ser. 4, Technology, Engineering and Safety, Vol. 2, Pergamon Press, Oxford (1960), p. 73.
[27] K. Sakodynskii and M. Lederer, *J. Chromatogr.* **20** (1965) 358.
[28] R. V. Davies, J. Kennedy, R. W. McElroy and K. M. Hill, *Nature* **203** (1964) 1110.
[29] G. Alberti, F. Dobici and G. Grassini, *J. Chromatogr.* **8** (1962) 103.
[30] K. A. Kraus, T. A. Carlson and J. S. Johnson, *Nature* **177** (1956) 1128.
[31] K. A. Kraus, H. O. Phillips, T. A. Carlson and J. S. Johnson, *Proc. Second Int. Conf. Peaceful Uses At. Energy*, Geneva (1958), Vol. 28, p. 3.
[32] G. Alberti and G. Grassini, *J. Chromatogr.* **4** (1960) 423.
[33] J. van R. Smit, *Nature* **181** (1958) 1530.
[34] M. Abe, *Bull. Chem. Soc. Jpn.* **42** (1969) 2683.
[35] G. Alberti, in *Ion Exchange*, edited by P. A. Williams and M. J. Hudson, Elsevier Applied Science, London (1987), p. 233.

[36] M. Lederer, *Anal. Chim. Acta* **12** (1955) 142.
[37] M. Lederer and S. Kertes, *Anal. Chim. Acta* **15** (1956) 226.
[38] G. H. Ayres, *Anal. Chem.* **25** (1953) 1622.
[39] S. K. Shukla, Doctoral Thesis, Paris (1962).
[40] V. Carunchio and G. Grassini, *Chromatogr. Rev.* **8** (1966) 260.
[41] M. Lederer, *J. Chromatogr.* **452** (1988) 265.
[42] M. Mazzei and M. Lederer, *J. Chromatogr.* **40** (1969) 197.
[43] S. Kaufman, *J. Am. Chem. Soc.* **82** (1960) 2963.
[44] R. G. Charles and S. Barnartt, *J. Inorg. Nucl. Chem.* **22** (1961) 69.
[45] S. Kaufman and L. S. Keyes, *Anal. Chem.* **36** (1964) 1777.

7 HPLC

This technique was first called high pressure liquid chromatography, then high performance liquid chromatography, and some call it high price liquid chromatography.

Its development solved two problems in liquid chromatography.

(i) A. J. P. Martin, while doing some separations on 50 cm long silica gel columns, realized that the efficiency of such columns was rather low because of the usual large particle size of the silica gel. When he tried to improve the efficiency by working with smaller particles, he found that the flow rate was so drastically reduced as to make the work tedious. He concluded (correctly) that to obtain good efficiency the small particles were essential, and sufficient pressure was needed to make the elution reasonably fast.

(ii) Cs. Horvath wanted to do some silica thin layer separations of the oxidation products of cholesterol. The solvent systems which gave good separations were very sensitive to atmospheric moisture and thus results varied from day to day.

As he had spent some years working on gas chromatography he felt that liquid chromatography should also be instrumentalized, and proceeded to build a liquid chromatograph along the lines of a gas chromatograph with sample injection, a detector and automatic recording of the detector response. Such a technique was no doubt needed in the modern laboratory, as is shown by the fact that about 160 000 such instruments are in use all over the world at the moment of writing (it may be more when this book is published).

The early designs of HPLC equipment were not adaptable to inorganic chromatographic problems. The columns and conduits and parts of the detector cell were made of stainless steel so as to with stand high pressures. However, eluents containing HCl or HBr or even some chlorides and bromides, attack stainless steel sufficiently to produce serious contamination. Some complexes, e.g. acetylacetone complexes, could be separated but, for example, Fe(acac)$_3$ (acac = acetylacetone) was found to react with finely divided silica and was in part irreversibly adsorbed.

On the other hand, excellent separations of *anions* were reported using ion pairing or anion exchange [1]. Figure 1 show the separations of some mixtures of anions using cetrimide as ion pairing agent on a cyano-bonded silica. Fig 9 in Chapter 12 on p. (168) shows the separation of condensed phosphates on an anion exchange resin.

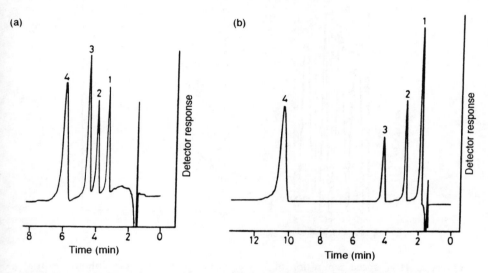

Figure 1. (a) Chromatogram of nitrogen anions and thiosulphate; 1 nitrite; 2, azide; 3, nitrate; 4, thiosulphate. (b) Chromatogram of halogen anions; 1, iodate; 2, bromate; 3, bromide; 4, iodide. (Reproduced by permission of Elsevier Science Publishers BV)

Techniques improved with time; notably, glass-lined columns have permitted the use of HPLC in a number of inorganic separations, of which we shall discuss some below.

Horwitz et al. [2] explored an obvious application of fast chromatography, the separation of short-lived radioelements. Figure 2 shows a separation of the ^{225}Ac series effected in 74 seconds with stepwise increase of the concentration of HNO$_3$ in the eluent.

Francium ($^{221}_{87}$Fr 4.8 min) has no long-lived or stable isotopes, thus HPLC offers a breakthrough for studying its chemistry.

Buckingham [3] reviewed his own work on the HPLC of Co(III) complexes. We shall quote two examples from his review:

THE ANATION OF [Co(en)$_2$(OH)$_2$)(OH)]$^{2+}$ BY C$_2$O$_4^{2-}$
This example illustrates the ability of RP HPIPC (reversed phase high performance ion pair chromatography) to detect and measure the concentrations of

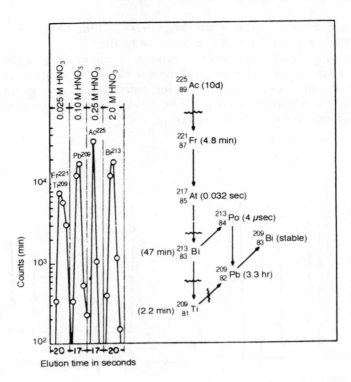

Figure 2. High-speed separation of ^{225}Ac and daughter nuclides. Column, Zorbax-SIL (5 μm), 10 mm × 2 mm i.d. with 30% (w/w) of di-(2-ethylhexyl) orthophosphoric acid in dodecane as stationary phase; eluent, HNO_3; flow rate, 17 cm/min; temperature, 50 °C [2]. (Reproduced by permission of Elsevier Science Publishers BV)

all species in a relatively complicated reaction sequence (Scheme 1). At pH ≈ 7 the anation occurs via the *cis* and *trans* hydroxo-aqua ions with little or no involvement of the diaqua and dihydroxo species. Ion pairing with $C_2O_4^{2-}$ is substantial, and is complete at $[C_2O_4^{2-}] > 0.3$ mol dm^{-3} for $[Co]_T \approx 10^{-3}$ mol dm^{-3}. Anation occurs via the ion paired species with both *cis*- and *trans*-$[Co(en)_2C_2O_4(OH_2)]^+$ intermediates being formed. Isolation of the *trans* intermediate allowed its chemistry to be studied independently; it isomerizes to the *cis*-monodentate which finally chelates via dual paths, displacement of water at the metal centre and attack by coordinated water at the carboxyl function. Apart from the ^{18}O-tracer work, all these processes can be followed chromatographically by quenching the reaction at appropriate times. Figure 3 shows an example of this under the conditions $[Co]_T = 8 \times 10^{-3}$ mol dm^{-3}, $[C_2O_4]_T = 0.3$ mol dm^{-3}, 25.0 °C and $I = 1.0$ mol dm^{-3}. The amounts of *cis*- and *trans*-$[Co(en)_2C_2O_4(OH_2)]^+$ and $[Co(en)_2C_2O_4]^+$ were monitored by quenching to pH ≈ 3 and freezing samples in liquid nitrogen until injection. The decay of the reactant ions and the growth of the two monodentate oxalato intermediates can be easily seen as can the slower growth of the final $[Co(en)_2C_2O_4]^+$ product. At higher pH values these reactions are reversed with $[Co(en)_2C_2O_4)_2]^+$ and *cis*-

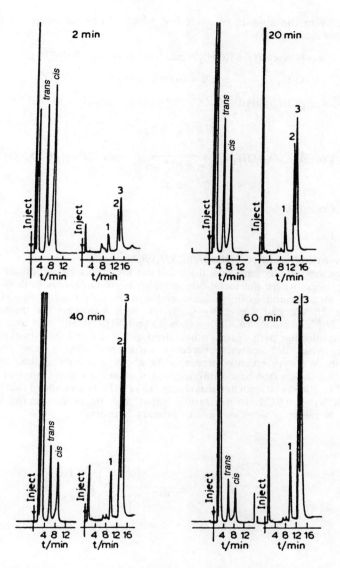

Figure 3. Samples (10 µl) of a reaction mixture consisting initially of $[Co(en)_2(OH_2)(OH)](ClO_4)_2$ (8×10^{-3} M), pH = 7.40 (0.1 M HEPES), 25.0 °C, $[C_2O_4^{2-}] = 0.3$ M. The *trans*-, *cis*-$[Co(en)_2(C_2O_4)(OH_2)]^+$ (1, 2) and $[Co(en_2(C_2O_4)]^+$ (3) ions were eluted with a 0–13.5% methanol gradient, 25 mM hexanesulphonate (pH 3.4); the *trans*- and *cis*-$[Co(en)_2(OH_2)]^{3+}$ ions were eluted with 40.5% methanol, 25 mM hexanesulphonate (pH 3.4) [3]. (Reproduced by permission of Elsevier Science Publishers BV)

and *trans*-$[Co(en)_2(C_2O_4)(OH)]$ giving rise to *cis*-$[Co(en)_2(OH)_2]^+$ and little or no *trans*-$[Co(en)_2(OH)_2]^+$. Clearly such processes are difficult, if not impossible, to quantify in the absence of a method which clearly distinguishes between alternative ionic species.

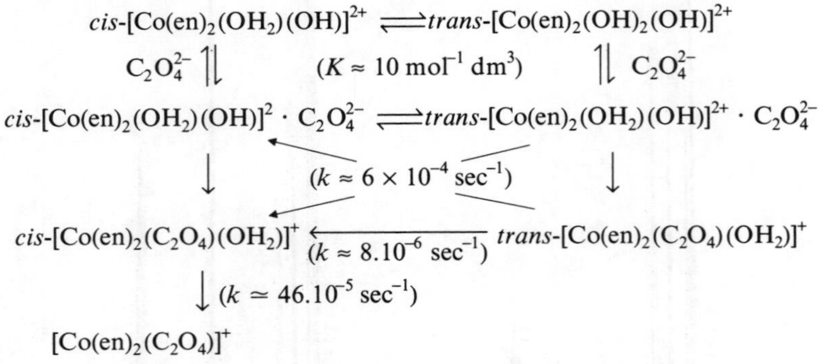
cis-[Co(en)2(OH2)(OH)]2+
trans-[Co(en)2(OH)2(OH)]2+
C2O42−
(K ≈ 10 mol−1 dm3)
C2O42−
cis-[Co(en)2(OH2)(OH)]2 · C2O42−
trans-[Co(en)2(OH2)(OH)]2+ · C2O42−
(k ≈ 6 × 10−4 sec−1)
cis-[Co(en)2(C2O4)(OH2)]+
(k ≈ 8.10−6 sec−1)
trans-[Co(en)2(C2O4)(OH2)]+
(k ≃ 46.10−5 sec−1)
[Co(en)2(C2O4)]+

Scheme 1

The base hydrolysis of the t-$[Co(tren)(NH_3)SCN]^{2+}$ ion

The question initially here was, does this ion undergo a base-induced isomerization process where the metal-bound thiocyanate undergoes both isomerization to metal-bound isothiocyanate and concomitant stereochemical change (Scheme 2). We had previously suggested such a process for *trans*-$[Co(en)_2NH_3(SCN)]^{2+}$, but only *ca.* 0.5% of *cis*-$[Co(en)_2NH_3(NCS)]^{2+}$ was accredited to the intramolecular path against a backdrop of 8% *cis*- and *trans*-isothiocyanate product. Also, the analysis procedure used previously is now considered uncertain. We have recently prepared the above *t*(SCN) complex, and other studies had shown that base-catalysed loss of an acido group from this position resulted in 80–90% stereochemical change to *p*(OH). It was hoped that substantial *t*(SCN) → *p*(NCS) isomerization might also occur (*t* signifies *trans* to tertiary N centre; *p* signifies *trans* to primary N centre).

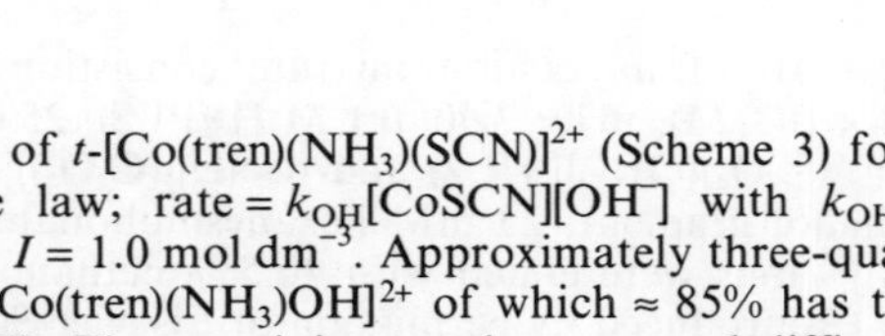
2+
N
NH2
NH2
Co
NH3
H2N
SCN
t(SCN)
(OH−)
2+
N
H2N
NH2
NCS
H2N
NH3
p(NCS)

Scheme 2

Base hydrolysis of *t*-$[Co(tren)(NH_3)(SCN)]^{2+}$ (Scheme 3) follows the expected second-order rate law; rate = $k_{OH}[CoSCN][OH^-]$ with $k_{OH} = 3.1 \times 10^{-2}$ mol^{-1} dm^3 s^{-1} at 25.0 °C, $I = 1.0$ mol dm^{-3}. Approximately three-quarters of the immediate product is $[Co(tren)(NH_3)OH]^{2+}$ of which ≈ 85% has the *p*-configuration and ≈ 15% is *t*(OH). The remaining products proved difficult to determine by conventional ion exchange chromatography but RP-HPIPC came to the rescue. Figure 4 shows chromatograms of the 1+ and 2+ products (*i.e.* 30% of the total) resulting from the hydrolysis of a 3 mg sample of *t*-$[Co(tren)(NH_3)$

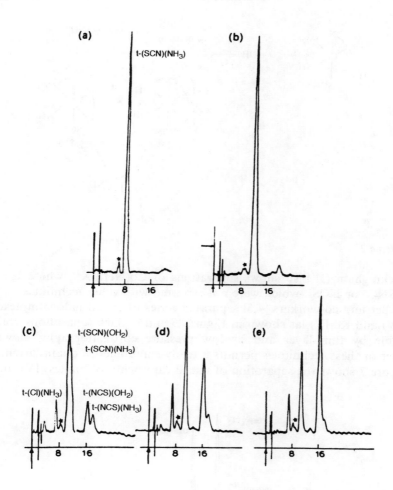

Figure 4. Samples (40 μl) following hydrolysis of 3.12 mg of t-[Co(tren)(NH$_3$) SCN]Br^2 in 0.4 cm^3 of 0.25 M sodium hydroxide at 25.0 °C for (a) immediate quench, (b) $t_{1/4}$ (45 s), (c) $2t_{1/2}$ (3 min), d $5t_{1/2}$ (7.5 min), (e) $10t_{1/2}$ (15 min). Aliquots were quenched into 0.5 M hydrochloric acid. (5% methanol, 25 mM toluenesulphonate (pH 3.5); flow rate 2 cm^3/min; a.u.f.s. 0.02; λ = 500 mm). (Reproduced by permission of Elsevier Science Publishers BV)

SCN]Br$_2$ in 0.4 cm^3 of 0.25 mol dm^{-3} sodium hydroxide at various times. Firstly, no p(NCS) is formed, but ≈ 5% isomerization to t(NCS) occurs (possibly via a tight ion-pair). However considerable hydrolysis of the p-NH$_3$ ligand occurs and the chromatograms show that both the t(SCN)p(OH) and t(NCS)p(NH$_3$) products subsequently hydrolyse to t(NCS)p(OH). The surprising result here is that t(SCN)p(OH) appears to isomerize totally to t(NCS)p(OH) under the alkaline conditions, and we are presently investigating this process.

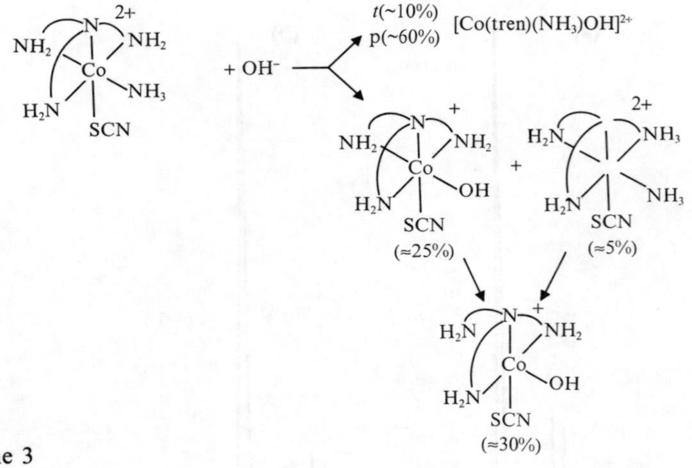

Scheme 3

Buckingham [3] also shows separations of $Co(NH_3)_5X^{2+}$, where $X = Cl^-$, Br^-, SNC^- or NCS^-, which were not reported with other techniques.

Müller and co-workers [4,5] separated series of mixed halo-complexes of Os(IV) and Re(IV), as shown in Figures 5 and 6. Such separations are also possible by thin layer and by low pressure chromatography; however, neither of these techniques permits a ready simultaneous quantitation.

Figure 7 shows the separation of mixed fluorochloroosmates (IV). In this

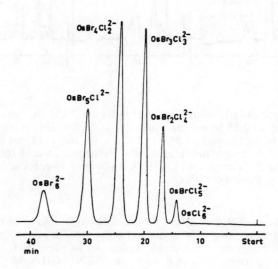

Figure 5. HPLC separation of all seven hexabromochloro-osmates(IV) on DEAE-silica gel with 0.5 M $HClO_4$ as eluent. (From [4])

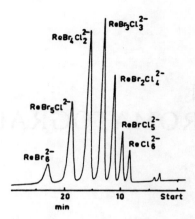

Figure 6. HPLC separation of all seven hexabromochlororhenates(IV) on DEAE-silica gel with 0.5 M HClO$_4$ as eluent. The small peaks near the start are hydrolysis products. (From [4])

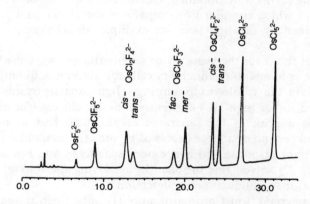

Figure 7. HPLC separation of mixed hexachlorofluoro-osmates(IV) on Nucleosil 10 Anion (Macherey–Nagel) with a gradient of aqueous HClO$_4$. The early peaks are hydrolysis products. (From [5])

series the isomers separate well, but this cannot be achieved with the mixed chlorobromo-complexes.

References

[1] R. N. Reeve, *J. Chromatogr.* **177**(1979) 393.
[2] E. P. Horwitz, W. H. Delphin, C. A. A. Bloomquist and G. F. Vandegrift, *J. Chromatogr.* **125**(1976) 203.
[3] D. A. Buckingham, *J. Chromatogr.* **313**(1984) 83.
[4] H. Müller and P. Bekk, *Fresenius' Z. anal. Chem.* **314**(1983) 758.
[5] P. Obergfell and H. Müller, *Fresenius' Z. anal. Chem.* **328**(1987) 242.

8 ION CHROMATOGRAPHY

In ion exchange column chromatography, excellent separations of alkalis or of the alkaline earths were obtained. Quantitation was effected by collecting exact volumes which contained one separated metal ion, and each ion was then determined by titration (see, for example, Beukenkamp and Rieman [1]).

I remember that over the years six or seven attempts were made to couple ion exchange columns to conductivity cells for automatic quantitation. The results were on the whole disappointing. High capacity resins were used, which needed rather high HCl concentrations for eluting (for example) the alkalis. Thus a peak of the separated alkali metal had to be measured against a background of a large excess of H^+ ions (of much higher mobility), resulting in poor response and hence poor sensitivity and low accuracy.

Small *et al.* [2] solved this problem by removing the ions of the eluting electrolyte prior to conductimetric detection.

In the commercial 'Ion Chromatograph' (Dionex Corp.), weak acid salts are used as eluents for anions. The eluates pass over the hydrogen form of a cation exchanger (called the 'suppressor column'), which removes the cation leaving only a weak acid as background. The zones of strong electrolytes can then be recorded with good sensitivity by a conductivity detector. Typical eluents are, e.g., carbonate–bicarbonate mixtures; as the resins used have a low capacity (0.007 meq/g), the concentration of the eluent can also be rather low (around 2×10^{-4}). Extensive work on a wide range of anions has been published, and some excellent separations are shown in Figures 1 and 2. Separations can be obtained in as little as 10 minutes.

Metal ions can be separated using, for example, dilute nitric acid and, as 'suppressor column', the OH^- form of an anion exchanger, so that the conductivity cell registers peaks of NaOH and KOH etc. against a low conductivity background (Figures 3 and 4).

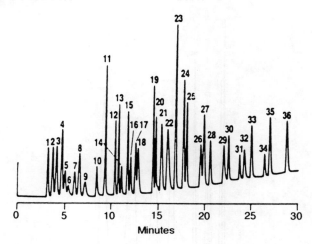

Figure 1. Gradient elution of inorganic and organic anions on a pellicular anion exchange resin. Eluent: gradient of 0.75–100 mM sodium hydroxide. Detection: suppressed conductimetric; anion micro-membrane suppressor. All anions 10 ppm unless noted otherwise. Peaks: 1 = F^- (1.5 ppm); 2 = α-hydroxybutyrate; 3 = acetate; 4 = glycolate; 5 = butyrate; 6 = gluconate; 7 = α-hydroxyvalerate; 8 = formate (5 ppm); 9 = valerate; 10 = pyruvate; 11 = monochloroacetate; 12 = BrO_3^-; 13 = Cl^- (3 ppm); 14 = galacturonate; 15 = NO_2^- (5 ppm); 16 = glucuronate; 17 = dichloroacetate; 18 = trifluoroacetate; 19 = HPO_3^{2-}; 20 = SeO_3^{2-}; 21 = Br^-; 22 = NO_3^-; 23 = SO_4^{2-}; 24 = oxalate; 25 = SeO_4^{2-}; 26 = α-ketoglutarate; 27 = fumarate; 28 = phthalate; 29 = oxalacetate; 30 = PO_4^{3-}; 31 = AsO_4^{3-}; 32 = CrO_4^{2-}; 33 = citrate; 34 = isocitrate; 35 = *cis*-aconitate; 36 = *trans*-aconitate. Courtesy Dionex Corporation. (From [3]; reproduced by permission of Elsevier Science Publishers BV)

Besides the 'suppressor columns' other methods have been developed. Among them is the 'membrane suppressor', which is a bundle of low density sulphonated polyethylene hollow fibres which permit the cations to permeate while the sulphonic groups repel the anions from the surface. The same effect as that of an ion exchange column is thus obtained.

Another alternative is simply to employ large organic anions, for example phthalic acid, as eluent; this yields a low conductivity background because large anions have lower conductivity than small ones.

Ion chromatography enjoys great popularity, firstly because it has solved various analytical problems, e.g. the trace analysis of anions and alkalis in environmental work, and because it was the first apparatus which offered the analyst the facilities usually associated with GC or HPLC: the injection of a sample leading to a record on a chart.

Five books on ion chromatography have appeared within a short time span: *Ion Chromatography* by J. S. Fritz, D. T. Gjerde and C. Pohlandt, Hüthig, Heidelberg, 1st edn. (1982); 2nd edn. (1987). *Ion Chromatography*, edited by J. G. Tarter, Marcel Dekker, New York (1987); *Ion Chromatography in*

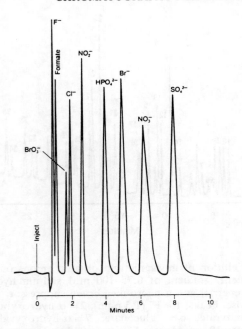

Figure 2. Separation of a nine anion standard on a HPIC-AS4 column with a 0.0028 M NaHCO$_3$–0.0022 M Na$_2$CO$_3$ eluent. Courtesy of Dionex Corporation. (From [4])

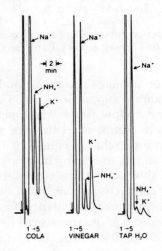

Figure 3. Separation of alkali metal ions in tap water [4]

Water Analysis, by O. A. Shpigun and Yu. A. Zolotov, Ellis Horwood, Chichester (1988); *Ion Chromatography* by P. R. Haddad and P. E. Jackson, Elsevier, Amsterdam (1990).

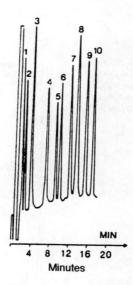

Figure 4. Separation of alkali, alkaline earth and transition metal ions [5]. Stationary phase: TSK-IC cation SW; mobile phase: 3.5 mM ethylenediamine-10.0 mM citric acid (pH = 2.8, adjusted with HCl); flow rate: 1.0 ml/min; detection: conductivity; temperature: 30 °C. Peaks: $1 = Na^+$ (5 ppm); $2 = K^+$ (50 ppm); $3 = Cu^{2+}$ (10 ppm); $4 = Ni^{2+}$ (10 ppm); $5 = Co^{2+}$ (10 ppm); $6 = Zn^{2+}$ (10 ppm); $7 = Fe^{2+}$ (10 ppm); $8 = Mn^{2+}$ (20 ppm); $9 = Cd^{2+}$ (10 ppm); $10 = Ca^{2+}$ (15 ppm). (Reproduced by permission of Elsevier Science Publishers BV)

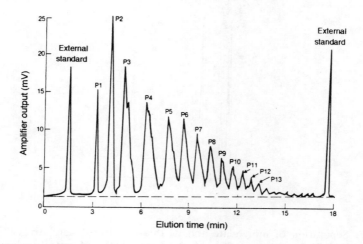

Figure 5. Neutralized polyphosphoric acid sample. Dashed line (---) indicates baseline and integration limits. P_1 indicates orthophosphate, P_2 indicates pyrophosphate, P_3 indicates triphosphate etc. External standard is 8.24 µg of P (as orthophosphate) per injection [4]

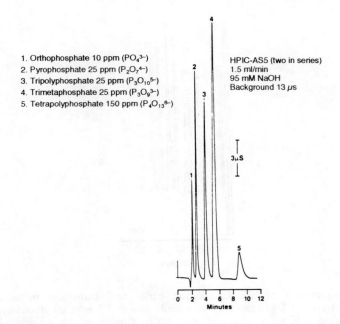

Figure 6. Separation of polyphosphates with a conductivity detector, using the anion membrane suppressor. (Reprinted from Ref. 6, p. 15 by courtesy of Marcel Dekker Inc.)

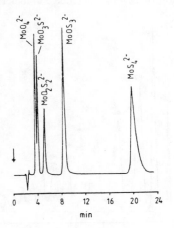

Figure 7. Separation of molybdate and its thia-substituted derivatives [7]. Separator: MPIC-NS 1; eluent: 0.001 M Na_2CO_3 + 0.002 M tetrabutylammonium hydroxide–acetonitrile (75:25 v/v); suppressed conductivity detection; flow rate: 1 ml/min; solute concentration: 50 mg/l $(NH_4)_2MoO_4$, 200 mg/l $(NH_4)_2MoO_2S_2$, $(NH_4)_2MoOS_3$ and $(NH_4)_2MoS_4$. (Reproduced by permission of Elsevier Science Publishers BV)

Although ion chromatography has been found very attractive for analyses, it does not offer much to the inorganic chemist as yet.

However, there are separations of polyphosphates (Figures 5 and 6) which are as good as those obtained by other techniques.

The most interesting application of this technique in inorganic chemistry is found in the paper by Weiss *et al.* [7] on the separation of thio- and seleno-molybdates and -tungstates. These compounds form when molybdate and tungstate solutions are treated with H_2S in concentrated ammonia

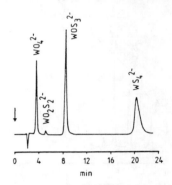

Figure 8. Separation of tungstate and its thia-substituted derivatives; solute concentration: 100 mg/l $Na_2WO_4 \cdot 2H_2O$ and Cs_2WOS_3, 200 mg/l $(NH_4)_2WS_4$. (Reproduced by permission of Elsevier Science Publishers BV)

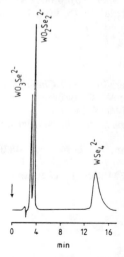

Figure 9. Separation of tungstate and its selena-substituted derivatives. Separator: MPIC-NS 1; eluent: 0.001 M Na_2CO_3 + 0.002 M tetrabutylammonium hydroxide–acetonitrile (62:38, v/v); UV detection at 254 nm; flow rate: 1ml/min; solute concentration: 100 mg/l $(NH_4)_2WO_2Se_2$ and 150 mg/l $(NH_4)_2WSe_4$. (Reproduced by permission of Elsevier Science Publishers BV)

solution. They also have importance in biochemistry, as they are formed in the digestive tract of ruminants and interfere with the normal copper metabolism. The separations shown below (Figures 7–10) were obtained with a Dionex 4020 liquid chromatograph.

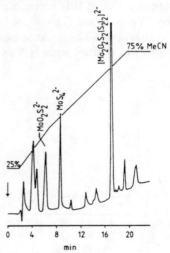

Figure 10. Gradient analysis of $[Mo_2O_2S_2(S_2)_2]^{2-}$ after hydrolysis of $MoO_2S_2^{2-}$. Separator: MPIC-NS 1; eluent (A) 0.001 M Na_2CO_3 + 0.002 M tetrabutylammonium hydroxide–acetonitrile (75:25, v/v), (b) 0.001 M Na_2CO_3 + 0.002 M tetrabutylammonium hydroxide–acetonitrile (25:75, v/v); UV detection at 254 nm; flow rate: 1 ml/min; reaction mixture diluted 1:10

References

[1] J. Beukenkamp and W. Rieman, *Anal. Chem.* **22**(1950) 582.
[2] H. Small, T. S. Stevens and W. C. Bauman, *Anal. Chem.* **47**(1975) 1801.
[3] H. Small, *J. Chromatogr.* *546*(1991) 11.
[4] see D.T. Gjerde and J.S. Fritz, *Ion Chromatography*, Verlag Hüthig, Heidelberg, 2nd edn (1987).
[5] S. Reiffenstuhl and G. Bonn, *J. Chromatogr.* **482**(1989) 289.
[6] J. G. Tarter, *Ion Chromatography*, Dekker, New York (1987), p. 15.
[7] J. Weiss, H. J. Möckel, A. Müller, E. Diemann and H. J. Walberg, *J. Chromatogr.* **439**(1988) 93.

9 GAS CHROMATOGRAPHY

In one of his early papers, A.J.P. Martin discussed the efficiency of chromatographic columns and pointed out that the rate of diffusion into and out of an adsorbent particle is one of the important factors in zone spreading. He then went on to mention that elution with a gaseous eluent should give highly efficient separations because diffusion in a gas is much faster than in a liquid. It was Martin who, with James, in the year 1952, developed the field of gas–liquid chromatography [1].

At that time gas–solid chromatography had already been described by some workers. We shall mention here the work of Janak in 1954, which is outstanding for its simplicity. He separated a number of gaseous compounds using CO_2 as eluent gas. The apparatus is shown in Figure 1. As the detector he used a gas pipette filled with NaOH solution, which absorbs all

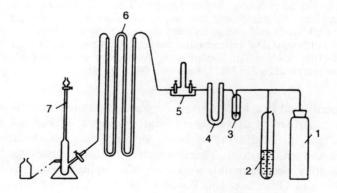

Figure 1. Gas chromatography as developed by Janak [2]. 1, CO_2 cylinder; 2, pressure regulator (mercury); 3, flow meter; 4, drying tube; 5, sample valve; 6, chromatographic column; 7, gas burette with NaOH solution (which removes the CO_2 used as eluent)

the CO_2, and thus a volume increase in the pipette is the volume of the chromatographed substance.

The original set-up used by James and Martin [3] is also worth mentioning. In their first experiments they separated the volatile fatty acids on a column of silica gel supporting paraffin oil as stationary phase. The issuing gas with the bands of the separated fatty acids was bubbled through a small beaker holding water with methyl red as pH indicator. With a burette, 0.1 N NaOH was added to maintain the indicator neutral. The 'chromatogram' consisted of a plot of ml of alkali consumed vs. time, as shown in Figure 2 below. The set-up required two operators: one to add the titrant and the other to write down the volume consumed. Soon afterwards Martin invented his gas density balance, which recorded density variations in the eluting gas stream.

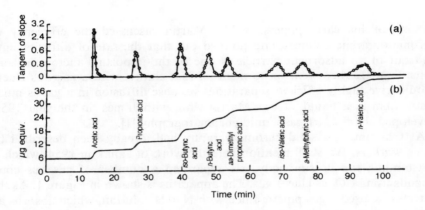

Figure 2. Gas–liquid partition chromatography. The separation of acetic, propionic, n-butyric and isobutyric acids and the isomers of valeric acid, showing the complete resolution of all bands and change in band shape in ascending series. (a) experimental curve; (b) differential of experimental curve. Column length, 11 ft.; liquid phase, stearic acid (10% w/w) in DC 550 silicone; nitrogen pressure, 74 cmHg; flow rate 18.2 ml/min; temperature, 137 °C (James and Martin [3])

Janak mentioned in a lecture some 20 years ago that about 50–60% of all analyses are performed using gas chromatography. It is thus not surprising that there is a splendid array of books dealing with gas chromatography techniques and applications. There are also several treatises on 'inorganic chromatography' dealing mainly with analytical applications of gas chromatography: those of Guiochon and Pommier [4], Moshier and Sievers [5], and Crompton [6].

In this book we shall not deal with the general principles of modern gas chromatography techniques as these are usually dealt with adequately in courses on analytical chemistry. Instead we shall mention only some topics

GAS CHROMATOGRAPHY

which are interesting in inorganic chemistry. For a good review of inorganic gas chromatography see also Uden [7].

Inorganic gases

There are numerous separations of such gases as O_2, N_2, CO_2, H_2S, SO_2, and NH_3. Several models of instruments for ultimate analysis use gas chromatography for determining CO_2, N_2, H_2O.

Figure 3 shows a separation of ppm amounts of COS, H_2S and SO_2 on a Teflon column with detection by a microwave-induced helium plasma for a specific elemental emission detection of sulphur gases [8].

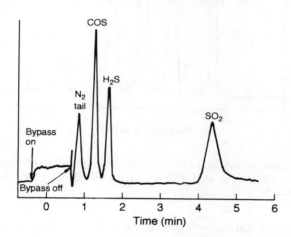

Figure 3. Chromatogram of 5 ppm each of COS, H_2S and SO_2 in N_2 [8]. Sample, 1 ml; column, 6 ft × 1/8 in. o.d. Teflon packed with Chromosil 310; temperature, 60 °C; flow rate, 15 ml/min He; total flow to plasma, 60 ml/min He; forward microwave power, 100 W; reflected power, 4 W; sulphur emission line, 182.04 nm. (Reproduced by permission of Elsevier Science Publishers BV)

HYDRIDES

Volatile hydrides such as silane, phosphine, arsine, stibine, germane and stannane have been chromatographed successfully. Figure 4 shows a separation of GeH_4, AsH_3, SnH_4 and SbH_3 carried out by Kadeg and Christian [9] using atomic absorption spectroscopy for detection.

Halides

There are some rather volatile inorganic halides, such as $GeCl_4$, which distils over from conc. HCl, and $HgCl_2$, which sublimes readily when heated in a test tube. However many other halides have been chromatographed at

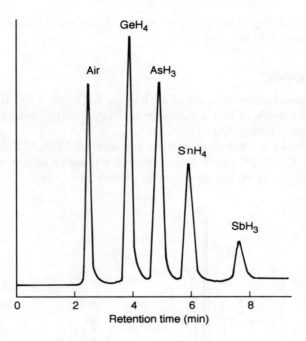

Figure 4. Separation of metalloid hydrides [9]. Column, 3 ft. Porapak Q; temperature programme 75–120 °C at 8 °C/min. (Reproduced by permission of Elsevier Science Publishers BV)

temperatures from 400–900°C or even up to 1500°C. We show here Figure 5 (from the review by Uden [7]).

Oxides

The oxides of Tc, Re, Os and Ir have been chromatographed at high temperatures. See Steffen and Bächmann [10].

Volatile metal complexes

We shall mention briefly a few examples which have found application in trace analysis.

MERCURY

Jones and Nickless [11] first used the Peters reaction with sodium benzenesulphonate to produce a phenylmercury salt which could be extracted and gas chromatographed. The analysis was quantitative down to 0.05 ppm. A still more sensitive reaction was to methylate inorganic mercury in aqueous

GAS CHROMATOGRAPHY

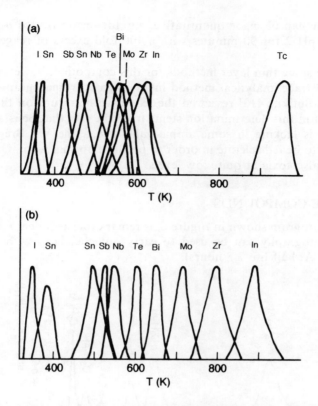

Figure 5. Separation of inorganic bromides [7]. Carrier gas, 60 mmHg of Br_2 and 4 mmHg of BBr_3 in N_2; stationary phases: (a) NaBr, (b) KBr on quartz granules; temperature programme, 400–900 °C. (Reproduced by permission of Elsevier Science Publishers BV)

solution by a trimethylsilyl salt [12] which, under suitable conditions, gave a quantitative conversion over the range 0.0025–10 ppm.

Good results have been obtained with sediment and fish samples.

SELENIUM

Selenium (IV) (i.e. selenite) reacts in acidic aqueous solution with *o*-phenylenediamine (and with substituted *o*-phenylenediamines) to yield *para*-diazaselenole (abbreviation: piazselenole), shown below:

The reaction can be made quantitative, e.g. for unsubstituted *o*-phenylenediamine at pH 2 for 90 minutes with a 100-fold excess of reagent over 20 μg of Se.

There are some thin layer methods for this compound; however, the most widely used trace analytical method involves gas chromatography.

Dilli and Sutikno [13] reviewed the extensive literature on this method. The interest in this determination stems from the fact that Se is an essential element. It is lacking in some alpine valleys (e.g. the Val Bregaglia) and sheep have to be fed selenite in order to produce normal wool. On the other hand, it is also toxic at quite low levels.

ASTATINE COMPOUNDS

The chromatogram shown in Figure 6, is remarkable as it demonstrates that gas chromatography can be used to carry out chemistry with short-lived isotopes (^{211}At half life 7.5 hours).

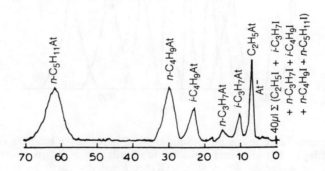

Figure 6. GLC separation of alkyl astatides, produced as a result of exchanging astatine, adsorbed at the column inlet in the form of astatines, with iodine in alkyl iodides. Column, 2 m × 4 m i.d., 10% dinonyl phthalate on Chromosorb G; carrier gas, He, flow rate 30 ml/min; column temperature, 95 °C [14]. (Reproduced by permission of Elsevier Science Publishers BV)

References

[1] A. T. James and A. J. P. Martin, *Analyst* **77**(1952) 915.
[2] J. Janak, *Collect. Czech. Chem. Commun.* **19**(1954) 684, 700, 917.
[3] A. T. James and A. J. P. Martin, *Biochem. J.* **50**(1952) 679.
[4] G. Guiochon and C. Pommier, *Gas Chromatography of Inorganics and Organometallics*, Ann Arbor Science Publishers, Ann Arbor, MI (1973).
[5] R. W. Moshier and R. E. Sievers, *Gas Chromatography of Metal Chelates*, Pergamon, Oxford (1965).
[6] T. R. Crompton, *Gas Chromatography of Organometallic Compounds*, Plenum Press, New York (1982).
[7] P. C. Uden, *J. Chromatogr.* **313**(1984) 3.

[8] J. L. Genna, W. D. McAninch and R. A. Reich, *J. Chromatogr.* **238**(1982) 103.
[9] R. D. Kadeg and G. D. Christian, *Anal. Chim. Acta* **88**(1977) 117.
[10] A. Steffen and K. Bächmann, *Talanta* **25**(1978) 51.
[11] P. Jones and G. Nickless, *J. Chromatogr.* **76**(1973) 285.
[12] P. Jones and G. Nickless, *J. Chromatogr.* **89** (1974) 201.
[13] S. Dilli and I. Sutikno, *J. Chromatogr.* **300** (1984) 265.
[14] M. Gesheva, A. Kolachkovsky and Yu. Norseyev, *J. Chromatogr.* **60**(1971) 414.

10 SEPARATION OF ISOTOPES

Attempts had already been made in 1937–1938 by Taylor and Urey [1, 2] to use chromatography for the separation of isotopes of Li, K and N using columns of artificial zeolites *35 metres* long. All they got was an enrichment. Later experiments by Glueckauf *et al.* [3] yielded an enrichment which was only a fraction of that expected from theory.

As the separation of isotopes is perhaps the extreme case for chromatography, it may be a good idea to consider the variables.

Isotopes have the same chemical properties but differ in mass. When the atomic number is small, e.g. with hydrogen–deuterium–tritium or with ^{6}Li–^{7}Li, the mass differences are large enough for a reasonable separation factor. However, a column of a sufficient plate number must be employed (a length of 35 metres sounds impressive but how many theoretical plates did it develop?). Also, in an aqueous solution the hydration of ions will tend to diminish the mass effect, viz ^{6}Li$^+$ will hydrate more than ^{7}Li$^+$.

The hydrogen isotopes

These isotopes were separated gas chromatographically by several workers. A typical separation employs an 80 metre capillary column at −196 °C with neon as carrier gas [4]. Bruner, Cartoni and Liberti separated numerous deuterated mixtures and even nitrogen isotopes by gas chromatography, as shown in Figures 1–3. For a discussion of the chromatography of labelled compounds, there is an excellent review by Klein [8], who shows that it is a current misconception that labelled compounds should behave exactly like the unlabelled ones. The separation factor may be small in most cases but this is not always so.

The separation ^{6}Li–^{7}Li

Ion exchange chromatography in aqueous solutions produces a change of the isotopic ratio along a chromatographed band. A typical result is shown

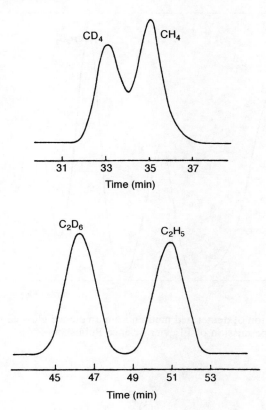

Figure 1. Chromatographic separation of CH_4–CD_4 and C_2H_6–C_2D_6 on a Porapak Q column (4.3 m × 0.20 mm i.d.). Ch_4–CD_4: T cal, 190 K; P_{N_2}, 1.5 atm; flow, 18.7 ml/min. C_2H_6–C_2D_6: T cal, 250 K; P_{N_2}, 1.75 atm; flow, 14.5 ml/min [5]. (Reproduced by permission of Elsevier Science Publishers BV)

in Figure 4. Similar, somewhat better, enrichments are obtained in paper electrophoresis also for ^{22}Na–^{24}Na (Figure 5) and even for ^{85}Rb–^{87}Rb. However, as mentioned above, much better separations are obtained in fused salts, in which the non-hydrated ions have bigger migration differences.

A preparative scale separation was developed by Klemm et al. [10] in 1947. A very simple apparatus used for the separation of lithium isotopes is shown schematically in Figure 6. The tube is made of Supremax glass and is filled with lithium chloride maintained at 650 °C. The U part of the apparatus is the separation tube and is 5.6 mm in diameter and filled with granular quartz of grain size 0.13 mm. Carbon electrodes and a 0.5 A current are used. During electrolysis, chlorine is liberated at the anode, and the metallic lithium deposited on the cathode is immediately converted to lithium chloride by a stream of chlorine. The counter-current is thus

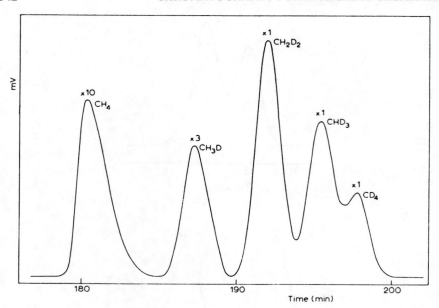

Figure 2. Separation of deuterated methanes on an etched glass capillary column [6]. (Reproduced by permission of Elsevier Science Publishers BV)

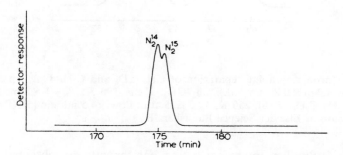

Figure 3. Separation of a mixture of $^{14}N_2$ and $^{15}N_2$ in an etched glass capillary column (175 m × 0.28 mm i.d.). Carrier gas, He–CO (45%); pressure, 13 cmHg; temperature, 77 K; flow rate, 0.6 ml/min [7]. (Reproduced by permission of Elsevier Science Publishers BV).

ensured, and this is one advantage of using fused salts. Furthermore, since the tube always contains pure lithium chloride, the conditions for total 'reflux' are always fulfilled; the liquid levels on the two sides of the separation tube are kept equal by regulating the nitrogen pressure on the cathode compartment. With such an apparatus, one gram of lithium chloride whose ^{6}Li content had altered from 7.3 to 16.1% was obtained after 4 days.

SEPARATION OF ISOTOPES

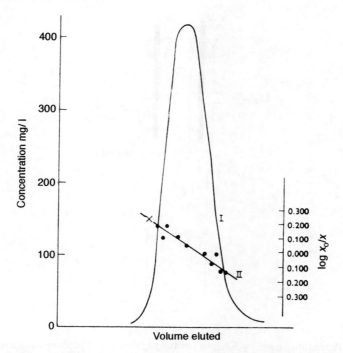

Figure 4. I: Elution curve of LiCl on Dowex-50; II: ratios ^{7}Li/^{6}Li expressed as $\log x_0/x$ [9]. (Reproduced by permission of Elsevier Science Publishers BV)

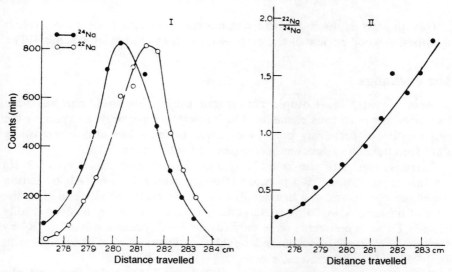

Figure 5. Electromigration of ^{22}Na–^{24}Na mixtures in aqueous solution. I: Separation of ^{22}Na and ^{24}Na; II: the ratio ^{22}Na/^{24}Na along the radioactive zone. (Reproduced by permission of Elsevier Science Publishers BV)

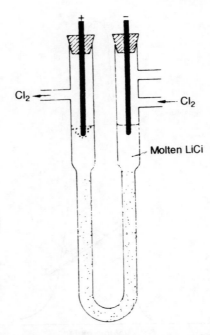

Figure 6. Apparatus used by Klemm for the separation of lithium isotopes. Chlorine is admitted at the cathode to prevent the deposition of metallic lithium. (Reproduced by permission of Elsevier Science Publishers BV)

This process is used for the commercial production of pure lithium isotopes. A good review of the early work is that written by Chemla [11].

Ion exchange

In recent years, the isotopic enrichment during ion exchange has been examined for numerous elements. The interest was not only in exploring the possibilities of preparing pure or enriched isotopes but also in providing data for elucidating geochemical cycles and interactions.

There is a series of papers by Kakihana, Kogure and co-workers [12–18] on this topic. Thus, ^{87}Rb is more strongly retained than ^{85}Rb by cation exchange resins, and the heavier isotope ^{7}Li is preferentially eluted in the case of lithium. Also, ^{41}K is preferentially retained while ^{39}K is more readily eluted. These experiments were carried out by displacing a band of K$^+$ by Ca^{2+} or Sr^{2+}. Strongly acidic and weakly acidic resins were examined, with both chloride and lactate as eluent.

Typical results are shown in Figures 7–9 and Table 1. In such work the method of determination of the isotope ration is usually mass spectrometry, requiring numerous mass spectra for each chromatogram. Thus, a single

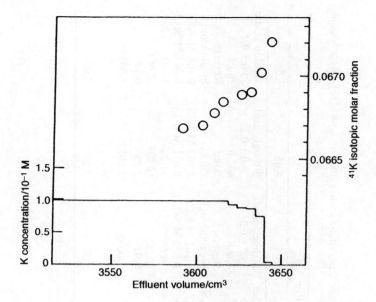

Figure 7. Chromatogram and the ^{41}K isotopic molar fractions in run K01 (Table 1). Experimental conditions are summarized in Table 1. The solid step-like line denotes the total potassium concentration and the open circles the ^{41}K isotopic molar fractions. The ^{41}K molar fraction of the feed solution is 0.06667 [18]. (Reproduced by permission of Elsevier Science Publishers BV)

chromatogram involves much work, and this is possibly the reason for the paucity of work in this interesting field.

SPEDDING'S SEPARATION OF NITROGEN ISOTOPES

Spedding et al. [19] developed a displacement method taking advantage also of the separation factor between dilute ammonium hydroxide and the ammonium form of Dowex 50–X12 (thus, the isotope effect between NH_3 and NH_4^+), which was found to be 1.0257 ± 0.0002. By displacing a rather long zone of ammonium ion with NaOH, it was shown that an adsorbed band of ammonium ion must be eluted through about 39 times its length in order to achieve a separation of the isotopes. They worked with 5 foot columns of Dowex 50 with a 4 in. diameter. Typical results are shown in Figure 10.

SEPARATION OF URANIUM ISOTOPES

From the work of Spedding, which separated $^{14}N-^{15}N$ (isotopes with a mass difference of 6.7%), it seems unlikely that uranium isotopes could be

Table 1. Experimental conditions and local enrichment factor and single-stage separation factor values obtained [18]. (Reproduced by permission of Elsevier Science Publishers BV)

Parameter	Run No.					
	K01	K02	K03	K04	K05	K06
Cation-exchange resin	Strongly acidic	Strongly acidic	Weakly acidic	Strongly acidic	Strongly acidic	Weakly acidic
Temperature (°C)	25.0 ± 0.2	25.0 ± 0.2	25.0 ± 0.2	25.0 ± 0.2	70.0 ± 0.2	70.0 ± 0.2
Resin bed height (cm)	410.7	404.5	202.5	408.6	405.0	200.4
Operating manner [band length cm]	Reverse breakthrough	Band (25.0)	Reverse breakthrough	Reverse breakthrough	Band (30.8)	Reverse breakthrough
Potassium feed solution[a]	0.101 M KCl	0.104 M KCl	0.102 M KCl	0.101 M KCl	0.103 M KCl	0.108 M KCl
Eluent[a]	0.049 M CaCl$_2$	0.050 M CaL$_2$	0.050 M CaCl$_2$	0.052 M SrCl$_2$	0.051 M SrL$_2$	0.048 M SrCl$_2$
Flow rate (cm^3 cm^{-2} h^{-1}) (cm^3 h^{-1})	11.51 (9.04)	9.87 (7.75)	13.72 (10.77)	10.65 (8.36)	11.34 (8.90)	11.03 (8.66)
Band velocity (cm h^{-1})	1.01	0.87	0.89	0.97	1.06	0.72
$R_{\text{local}}^{\max}$	1.0085	1.0065	1.0097	1.0092	1.0027	1.0039
$R_{\text{local}}^{\min}$	–	0.9953	–	–	0.9954	–
$10^5 \varepsilon$	2.52	2.35[b]	3.48	2.52	1.25[c]	2.49

[a] L = lactate.
[b] Average of values from the front part (2.64×10^{-5}) and the rear part (2.07×10^{-5}) of the band.
[c] Average of values from the front part (1.33×10^{-5}) and the rear part (1.16×10^{-5}) of the band.

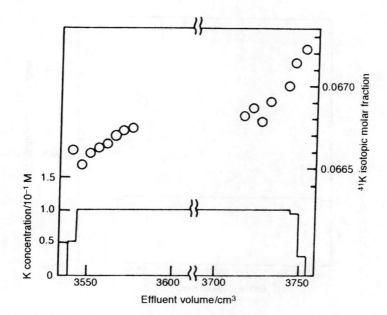

Figure 8. Chromatogram and the ^{41}K isotopic molar fractions in run K02. Experimental conditions are summarized in Table 1. The solid step-like line denotes the total potassium concentration and the open circles the ^{41}K isotopic molar fractions. The ^{41}K molar fraction of the feed solution is 0.06682 [18]. (Reproduced by permission of Elsevier Science Publishers BV)

separated, as the mass difference for the separation of ^{235}U–^{238}U is only 1.3%; however, this is not so at all. The new edition of 'Gmelin' [20] devotes a large chapter to such separations, written by Y. Marcus, and he sums up the situation in his introduction as follows.

> The separation of the isotopes of uranium, and specifically the enrichment of the ^{235}U content in uranium from its natural level of 0.71% to that useful for nuclear reactors, 3%, is of very high significance in nuclear technology. The methods commonly used, i.e., gaseous diffusion and gas centrifugation of UF_6, are complicated, expensive, and demanding of energy. In the search for technologies that are simpler, cheaper, and that require less energy, chemical exchange methods have been the subject of extensive investigations...... Amongst these methods, those involving ion exchange resins and membranes have constituted a major fraction of the efforts.
>
> The use of ion exchangers for the separation of isotopes started with a demonstration of an isotope effect in the exchange of lithium ions between a solution and an inorganic ion exchanger by Taylor and Urey. Attempts to separate the isotopes of uranium by a similar method followed during and immediately after the Manhattan Project in the USA in the 1940's. The earlier results were inconclusive, but subsequent work at the Oak Ridge National

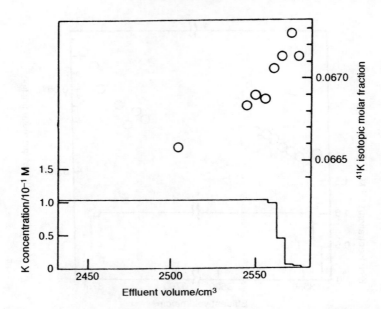

Figure 9. Chromatogram and the ^{41}K isotopic molar fractions in run K03. Experimental conditions are summarized in Table 1. The solid step-like line denotes the total potassium concentration and the open circles the ^{41}K isotopic molar fractions. The ^{41}K molar fraction of the feed solution is 0.06667 [18]. (Reproduced by permission of Elsevier Science Publishers BV)

Laboratory and at the Ames Laboratory, which followed at the early 1950's, was more definite. The investigators were optimistic concerning the feasibility of separating the isotopes of uranium economically by means of ion exchange reactions, in view of the surprisingly large single stage separation factors $\varepsilon = 20 \times 10^{-4}$ that they have obtained. This is not so far from the square root of the ratio of the isotopic masses minus one, 56×10^{-4}, which is the limiting value theoretically expected for the bare ions. These large experimental values must, however, have been due partly to faulty assays, since subsequent work on the same systems has consistently resulted in much lower values, at best $\varepsilon = 0.5 \times 10^{-4}$. Such a small value casts strong doubts on the technical feasibility and the economics of the separation of the isotopes of uranium by means of ion exchange.

Nevertheless, several Japanese groups, led by Kakihana, who had done pioneering work in this field have persevered in developing this method during the last 20 years. According to their later findings (see, e.g., [18]), $\varepsilon = 2 \times 10^{-4}$ is attainable in certain systems, and this can lead to the production of 8 kg of 3% enriched ^{235}U per m^3 resin per year, which they evaluated to be economically attractive. Higher values of ε have been obtained in certain other systems, too. Groups elsewhere, especially a Roumanian one led by Călușaru, who started his work in the 1970's, and still others also took up the challenge and continue the work on this problem.

There are three general methods involved in the work on the separation of the isotopes of uranium by ion exchange. One involves the breakthrough or the

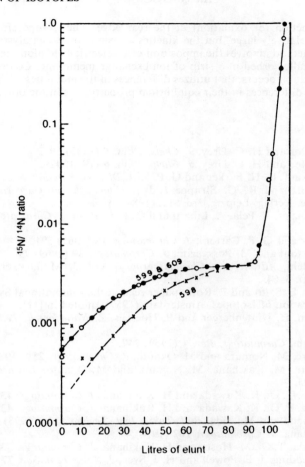

Figure 10. Profiles of the adsorbed ammonium band on a series of 4 in. × 59 in., 100–200 mesh, Dowex 50-X12 resin beds. The adsorbed band was eluted in a semi-continuous manner by adding the equivalent of 550 ml of 15 N ordinary NH_4OH and withdrawing the same amount of NH_4OH depleted in ^{15}N every circuit of the 10 column series. Profile 598 was taken just prior to, and profiles 599 and 609 just after, an addition and a withdrawal. Profiles 599 and 609 are exactly one cycle apart and are virtually identical [19]. (Reprinted with permission from Spedding *et al.* [19]. Copyright (1955) American Chemical Society)

displacement of a band of either U^{IV} or U^{VI} along an ion exchange column, where the isotope fractionation is due to differences in the affinities of ^{235}U and ^{238}U species to the resin relative to the solution. The isotope effect is small, but the kinetics are reasonably fast, if diffusion in the resin is not hampered by too tight cross-linking. Another method involves the isotope exchange between U^{IV} and U^{VI}, one of these valence states being sorbed on the resin and the other being predominantly in the solution. A uranium band is moved along the resin column with oxidation (or reduction) occurring at the front edge of the band

and/or reduction (or oxidation) at the rear edge. The isotope effect of this process is relatively large, but the kinetics are slow, unless catalysts are used. The third method involves the imposition of an electric field along the uranium band, generally sorbed in a strip of ion exchange membrane. Counter-current electromigration occurs, that utilizes differences in the mobilities of the isotopes rather than differences in their equilibrium properties, as in the other methods.

References

[1] T. I. Taylor and H. C. Urey, *J. Chem. Phys.* **5** (1937) 597.
[2] T. I. Taylor and H. C. Urey, *J. Chem. Phys.* **6** (1938) 429.
[3] E. Glueckauf, K. H. Barker and G. P. Kitt, *Discuss. Faraday. Soc.* **7** (1949) 199.
[4] E. Leibnitz and R. G. Struppe (Eds.), *Handbuch der Gaschromatographie*, Akademie. Verlag, Leipzig, 3rd edn. (1984), 310.
[5] M. Possanzini, A. Pela, A. Liberti and G. P. Cartoni, *J. Chromatogr.* **38** (1968) 492.
[6] F. Bruner and G. P. Cartoni, *J. Chromatogr.* **18** (1965) 391.
[7] G. P. Cartoni and M. Possanzini, *J. Chromatogr.* **39** (1969) 99.
[8] P. D. Klein, *Advances in Chromatography*, Vol. 3, M. Dekker, New York (1966), pp. 3–64.
[9] F. Menès, E. Saïto and E. Roth, paper read at the International Symposium on the Separation of Isotopes, Amsterdam (1957), quoted in [11].
[10] A. Klemm, H. Hintenberger and P. Hoernes, *Z. Naturforsch. Teil A* **2a** (1947) 245.
[11] M. Chemla, *Chromatogr. Rev.* **1** (1959) 247.
[12] K. Kogure, M. Nomura and M. Okaoto, *J. Chromatogr.* **259** (1983) 480.
[13] K. Kogure, M. Kakihana, M. Nomura and M. Okaoto, *J. Chromatogr.* **325** (1985) 195.
[14] M. Hosoe, T. Oi, K. Kawada and H. Kakihana, *J. Chromatogr.* **435** (1988) 253.
[15] M. Hosoe, T. Oi, K. Kawada and H. Kakihana, *J. Chromatogr.* **438** (1988) 225.
[16] H. Kakihana and M. Aida, *Bull. Tokyo Inst. Technol.* **116** (1973) 39.
[17] H. Kakihana, *J. Chromatogr.* **102** (1974) 47.
[18] K. Kawada, T. Oi, M. Hosoe and H. Kakihana, *J. Chromatogr.* **538** (1991) 355.
[19] F. H. Spedding, J. E. Powell and H. J. Svec, *J. Am. Chem. Soc.* **77** (1955) 6125.
[20] Y. Marcus, in *Gmelin Handbook of Inorganic Chemistry*, 8th edn, Uranium, Suppl. Vol. 4D, Springer, Berlin (1983), p. 204.

11 SEPARATION OF OPTICAL ISOMERS

Until relatively recently, there existed only the three methods of Pasteur (developed in 1860) for the separation of optical isomers; namely, the separation of assymetric crystals under the microscope, the attack of one isomer by suitably selective bacteria, and the precipitation of one isomer with a suitable optically active precipitant, e.g. l-strychnine for acids or d-bromocamphorsulphonic acid for bases.

The separation of optical isomers of inorganic complexes has played a role in complex chemistry, as it provided a method for establishing the octahedral structure of hexa-coordinated complexes, e.g. the Co(III) or Cr(III) complexes.

It also permitted the establishment of the *trans* or *cis* positions of isomeric Co(III) complexes because

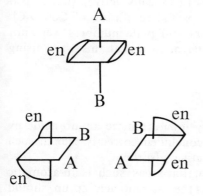

has no optical isomers, whereas

does have optical isomers.

Finally Alfred Werner in 1914 resolved a completely inorganic complex into its d- and l-forms

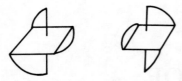

where the bidentate ligand was

$$\begin{array}{c} NH_3 \\ HO \\ HO \\ NH_3 \\ NH_3 \end{array}$$

The polynuclear compound was obtained by the action of ammonia on chloroaquotetramminecobalt(III) ion $[Co(NH_3)_4Cl\,H_2O]^{2+}$. It has a very high rotation, $[M]_d = \pm 47\,600°$. This compound refuted the previously held belief that optical activity must always be associated with organic (i.e. living) substances.

The Nobel prize was given to Werner (1866–1919) 'in recognition of his work on the linkage of atoms in molecules by which he has thrown fresh light on old problems and opened up new fields of research, especially in inorganic chemistry'.

Early chromatographic experiments on complexes were carried out by Karagunis and Coumoulos [1], who obtained partial separations of $[Cr(en)_3]Cl_3 \cdot 3.5\,H_2O$ isomers on a column of (optically active) quartz, also by Krebs and Rasche [2], who obtained a partial resolution of $Co(en)_3Cl_3$ on starch. In both cases separation factors and plate numbers were not seriously considered, and it may be of interest to repeat this work using modern concepts.

Dalgliesh's 'three point' theory

Dalgliesh [3] obtained separations of (very few) aromatic amino acids by partition chromatography on paper, and concluded correctly that in such separations cellulose acts as an optically active adsorbent.

He pointed out that substances containing flat areas, such as an aromatic ring, and hydrogen bonding groups, such as the α-amino acid group, should be capable of being adsorbed on the long flat molecules of cellulose, which contain large numbers of groups forming hydrogen bonds. A 'three point' attachment was postulated: at the α-amino group, the carboxyl and one substituent in the ring allow a closer fit with the cellulose surface and hence greater adsorption of one of the antipodes.

This picture seems essentially correct. Later work seems to suggest that such chiral adsorption takes place mainly in the amorphous regions of the cellulose, i.e. possibly in cone-shaped pockets.

The adsorption, in order to produce a good chromatogram, should involve three relatively weak interactions, otherwise both isomers are too strongly adsorbed. The slower spot, i.e. the more adsorbed, where all 'three points' interact will usually be elongated, because kinetics involving 'three points' are slower than those involving single point or two point attachment.

For the separation of octahedral Co(III) complexes, a similar 'three point' model can be postulated for the formation of outer sphere complexes with optically active counter-ions, viz d- or l-tartrate, as well as some complex tartrates such as antimony tartrate and arsenic tartrate, l-strychnine (i.e. the optically active isomer occurring naturally) and d-bromocamphorsulphonate.

Successful chromatographic separations have been effected by either linking the chiral agent, e.g. tartrate, to the support or by eluting with an eluent which contains an optically active counter-ion.

Electrophoretic separations were also successful with d- or l-tartrate or complex tartrates as electrolytes for cationic complexes and l-strychnine sulphate for anionic complexes. Here the chiral discrimination must be sufficiently large at an intermediate range of concentrations, so that the ions may migrate at reasonable speeds in an environment that is not too dilute (which would cause tail formation) and not too concentrated (otherwise there is too little movement).

We shall now discuss some examples of successful separations of optical isomers of metal complexes.

Figure 1 shows the separation of d- and l-$[Co(en)_3]^{3+}$ eluted by 0.15 M sodium d-tartrate on an SE-Sephadex column (with sulphonic groups) by Yoshikawa and Yamasaki [4].

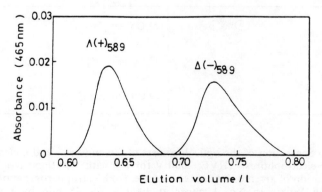

Figure 1. Elution curve of $[Co(en)_3]^{3+}$ eluted by 0.15 M sodium d-tartrate on an SE-Sephadex column [6]. (Reproduced by permission of Elsevier Science Publishers BV)

Fujita *et al.* [5] prepared a cation exchanger suitable for HPLC by allowing Toyopearl HW-40 a polyvinyl gel with hydroxy groups like Sephadex) to react with D-tartaric acid, i.e. a resin with D-tartaric acid groups. A separation of $[Co(tn)_3]^{3+}$ isomers with aqueous L-tartrate as eluent (and with recycling) is shown in Figure 2. There is a double effect here: both support and eluent are optically active ion pairing agents.

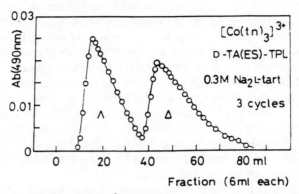

Figure 2. Elution curve of $[Co(tn)_3]^{3+}$ on a D-TA(ES)-Toyopearl column [6]. (Reproduced by permission of Elsevier Science Publishers BV)

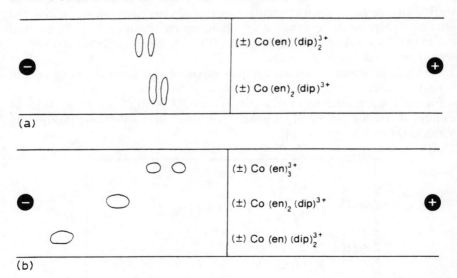

Figure 3. (a) Separation of the optical isomers of $Co(en)(dip)_2^{3+}$ and $Co(en)_2(dip)^{3+}$ by high voltage electrophoresis (HVE; 2000 V for 1 h with Whatman No. 1 paper) in a 0.25 M solution of arsenic sodium (+)-tartrate. (b) Electropherogram of cobalt(III) complexes on Whatman No. 1 paper at 3000 V for 45 min by HVE (Camag apparatus) in 0.2 M antimony potassium (+)-tartrate [8]. (Reproduced by permission of Elsevier Science Publishers BV)

Antimony d-tartrate [it is actually (bis-μ-d-tartrato) diantimonate(III), a divalent anion with a dimeric structure] is usually better than d-tartrate as an ion pairing agent, and hence also as eluent in ion exchange separations. Yoneda [6] gives examples of the better ion exchange and reversed phase separations achieved by its use. Even non-charged complexes such as fac-[Co(β–ala)$_3$] tend to separate into their isomers on ion exchange columns with tartrate and antimony tartrate as eluent. Yamazaki et al. [7] discuss the mechanism of such separations, which seem to be based on ion pairing with the eluent.

The [Co(en)$_3$]$^{3+}$ isomers separate well in paper electrophoresis with l- and in d-tartarate, but mixed dip–en or O-phen–en complexes separate only in antimonyl tartrate or arsenic tartrate, which has, as mentioned above, better chiral discrimination. See Figure 3 and 4. The importance of the pH and the tartrate concentration is illustrated in Figure 5.

Figure 6 shows what happens when the optically active electrolyte is not 'optically pure', i.e. if the l-tartrate employed contains some d-tartrate.

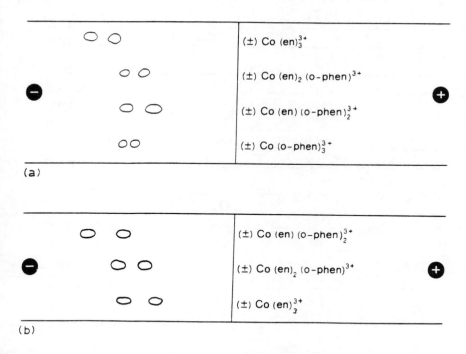

Figure 4. (a) Separation of the optical isomers of cobalt(III) complexes with mixed ligands by HVE (2500 V for 1 h with Whatman No. 1 paper) in 0.3 M arsenic sodium (+)-tartrate. (b) Electropherogram of cobalt(III) complexes with mixed ligands on Whatman No. 1 paper at 3000 V for 45 min in 0.2 M antimony potassium (+)-tartrate (Camag HVE apparatus) [8]. (Reproduced by permission of Elsevier Science Publishers BV)

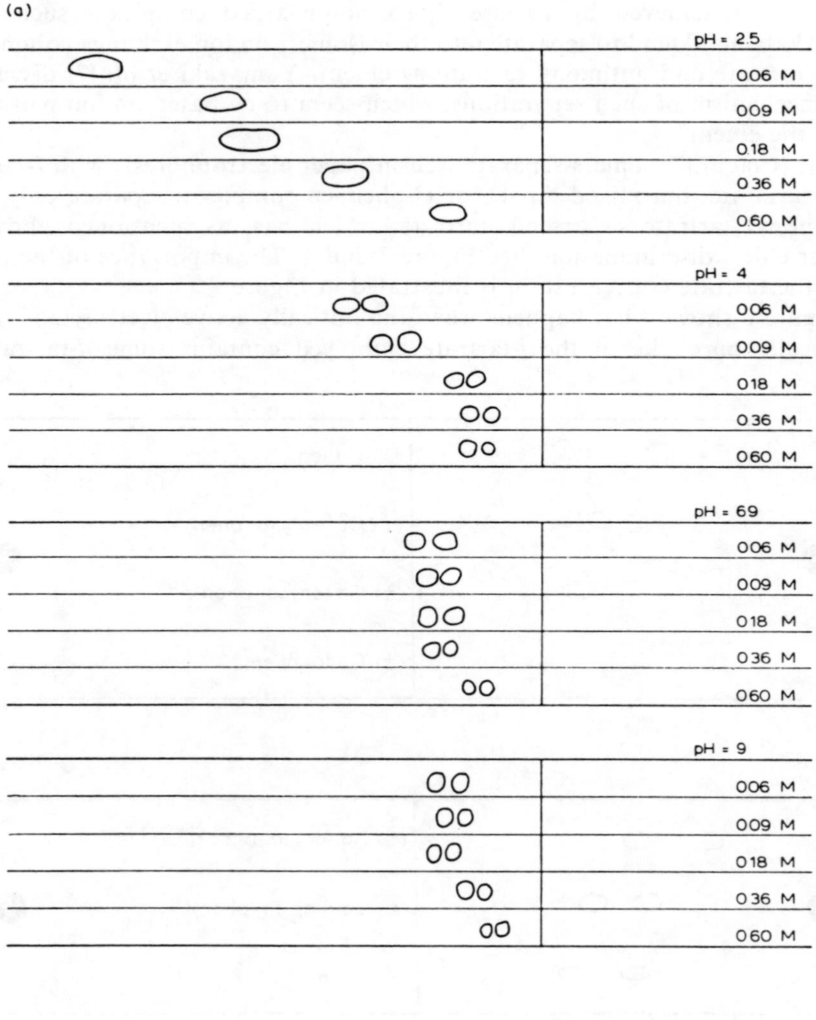

Figure 5. Electropherograms of cobalt(III) complexes on Whatman No. 1 paper in d-tartrate solutions at 1500 V for 1 h at 8 °C. (a) Racemic $(Co(en)_3^{3+}$ at various pH values and concentrations; (b) $Co(dip)_3^{3+}$, $Co(o\text{-phen})_3^{3+}$ and $Co(en)_3^{3+}$ placed side by side in various concentrations of d-tartrate at pH 6.9 [8]. (Reproduced by permission of Elsevier Science Publishers BV)

SEPARATION OF OPTICAL ISOMERS

(b)

	d-tartrate 0.06 M
○	Co(dip)$_3^{3+}$
○	Co(o-phen)$_3^{3+}$
○○	Co(en)$_3^{3+}$

	0.09 M
○	Co(dip)$_3^{3+}$
○	Co(o-phen)$_3^{3+}$
○○	Co(en)$_3^{3+}$

	0.18 M
○	Co(dip)$_3^{3+}$
○	Co(o-phen)$_3^{3+}$
○○	Co(en)$_3^{3+}$

	0.36 M
○	Co(dip)$_3^{3+}$
○	Co(o-phen)$_3^{3+}$
○○	Co(en)$_3^{3+}$

	0.60 M
○	Co(dip)$_3^{3+}$
○	Co(o-phen)$_3^{3+}$
○○	Co(en)$_3^{3+}$

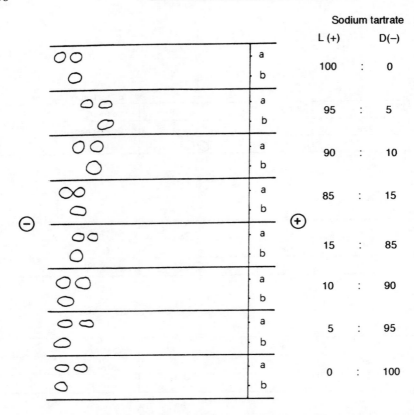

Figure 6. High voltage electrophoresis of $Co(en)_3^{3+}$ complexes on Whatman No. 1 paper at 2500 V for 35 min at 6 °C. Electrolytes: mixtures of 0.18 M sodium L(+)- and D(−)-tartrate. (a) Mixture of (+)-$Co(en)_3^{3+}$ and (−)-$Co(en)_3^{3+}$; (b) (+)-$Co(en)_3^{3+}$ [9]. (Reproduced by permission of Elsevier Science Publishers BV)

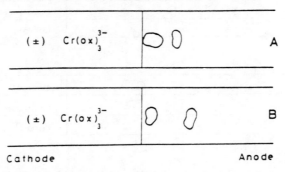

Figure 7. Separation of the optical isomers of $Cr(ox)_3^{3-}$ using acetic acid containing 15% (w/v) of brucine as electrolyte. Whatman No. 1 paper at 2000 V for $1\frac{1}{4}$ h. Acetic acid concentration: (A) 0.2 N; (B) 0.3 N [10]. (Reproduced by permission of Elsevier Science Publishers BV)

Although the distance between the two zones diminishes, the separation still yields two 'optically pure' spots[9].

Anionic complexes such as $[Co(ox)_3]^{3-}$ or $[Cr(ox)_3]^{3-}$ can be resolved with *l*-brucine or with quinine as shown in Figures 7–9. Another example of good separations of optical isomers is shown by the separation of the tris[*trans*-cyclohexane-1, 2-diamine]Co(III)$^{3+}$ complexes in Figures 10 and 11.

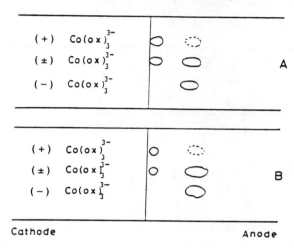

Figure 8. Separation of the optical isomers of $Co(ox)_3^{3-}$ using Whatman No. 1 paper at 2000 V for $1\frac{1}{2}$ h with the following electrolytes: (A) 0.3 N acetic acid containing 10% (w/v) of quinine; (B) 0.3 N acetic acid containing 20% (w/v) of quinine [10]. (Reproduced by permission of Elsevier Science Publishers BV)

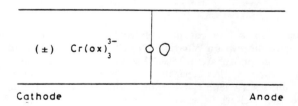

Figure 9. Separation of the optical isomers of $Cr(ox)_3^{3-}$ by high voltage electrophoresis (2000 V for $1\frac{1}{2}$ h with Whatman No. 1 paper) in 0.3 N acetic acid containing 20% (w/v) of quinine as electrolyte [10]. (Reproduced by permission of Elsevier Science Publishers BV)

For further reading on this topic, I would recommend H. J. Eméleus and J. S. Anderson, *Modern Aspects of Inorganic Chemistry*, Routledge, London (1942) and the review by Yoneda entitled 'Mechanism of chromatographic separation of optically active metal complexes'[6].

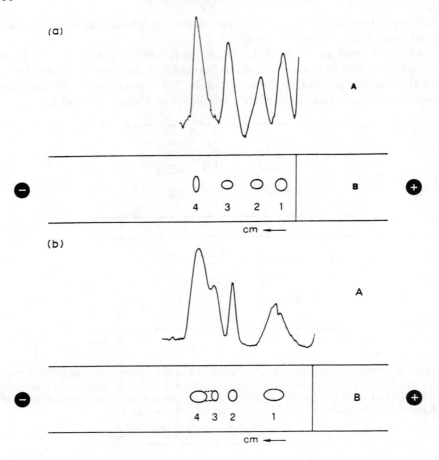

Figure 10. Separation of four catoptromer complexes of Cobalt(III)(±)-(chxn)$_3^{3+}$ by high voltage electrophoresis on Whatman No., 1 paper at 1500 V for 60 min. (a) 0.2 M solution of sodium phosphate; (b) 0.2 M solution of sodium sulphate. B = Electropherogram; A = scan of the separate isomers in electropherogram B. Wavelength = 450 nm; slit length = 4 mm; slit width = 0.1 mm; V_p (plate travel) = V_c(recorder chart speed) = 1 mm/s. 1 = Λ + Δ lel$_3$; 2 = Λ + Δ lel$_2$ob; 3 = Λ + Δ lel ob$_2$; 4 = Λ + Δ ob$_3$ [11]. (Reproduced by permission of Elsevier Science Publishers BV)

References

[1] G. Karagunis and G. Coumoulos, *Nature* **142** (1938) 162.
[2] H. Krebs and R. Rasche, *Naturwissenschaften* **41** (1954) 63.
[3] C. E. Dalgliesh, *J. Chem. Soc.* (1952), 3940.
[4] Y. Yoshikawa and K. Yamasaki, *Inorg. Nucl. Chem. Lett.* **6** (1970) 523.
[5] M. Fujita, Y. Yoshikawa and H. Yamatera, *Chem. Lett.* (1982) 437.
[6] H. Yoneda, *J. Chromatogr.* **313** (1984) 71.

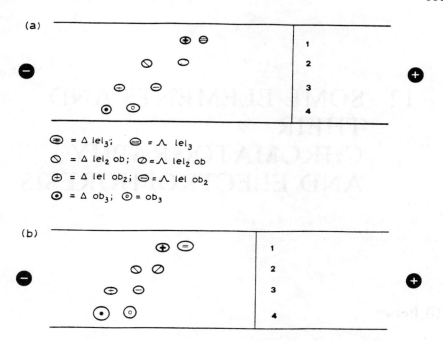

Figure 11. Resolution of Λ and Δ forms of the four catoptromer complexes of cobalt(III)–*trans*-cyclohexane-1,2-diamine on Whatman No. 1 paper at 2000 V for 45 min by HVE (Camag apparatus). (a) Sodium (+)-tartrate, 0.2 M, pH = 6.8; (b) ammonium (+)-tartrate, 0.2 M, pH 6.8 [11]. (Reproduced by permission of Elsevier Science Publishers BV)

[7] S. Yamazaki, T. Yukimoto and H. Yoneda, *J. Chromatogr.* **175** (1979) 317.
[8] L. Ossicini and C. Celli, *J. Chromatogr.* **115** (1975) 655.
[9] S. Fanali, M. Lederer, P. Masia and L. Ossicini, *J. Chromatogr.* **440** (1988) 361.
[10] V. Cardaci and L. Ossicini, *J. Chromatogr.* **198** (1980) 76.
[11] S. Fanali, L. Ossicini and T. Prosperi, *J. Chromatogr.* **318** (1985) 440.

12 SOME ELEMENTS AND THEIR CHROMATOGRAPHY AND ELECTROPHORESIS

(i) Boron

BORIC ACID AND POLYBORATES

Guedes de Carvalho [1] has published extensive studies on the paper chromatography of boric acid and borates at different pH values. There seems to be only one spot formed, indicating very fast equilibrium between the various polymeric forms.

HALOGENOBORATES

Decahydroborate $B_{10}H_{10}^{2-}$ reacts with halogens to yield halogenohydroborates, which can be separated by high voltage electrophoresis [2, 3]. Figures 1–3 show the separations obtained with chlorohydroborates, bromo-

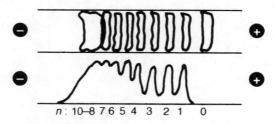

Figure 1. Electrophoretic separation of ^{36}Cl-labelled chlorohydroborates $B_{10}H_{10-n}Cl_n^{2-}$ ($n = 1$–10). On Schleicher & Schüll Bmgl paper with 0.2 M CCl_3COOH–KOH (pH 2.5) as electrolyte with 60 V/cm at 1–5 °C for 1–3 hours. (Reproduced by permission of Elsevier Science Publishers BV)

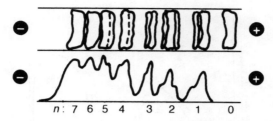

Figure 2. Electrophoretic separation of ^{82}Br-labelled bromohydroborates $B_{10}H_{10-n}Br_n^{2-}$. On Schleicher & Schüll Bmgl paper with the same conditions as in Figure 1

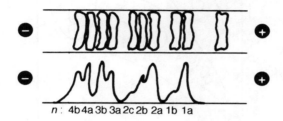

Figure 3. Electrophoretic separation of ^{131}I-labelled iodohydroborates $B_{10}H_{10-n}I_n^{2-}$ ($n = 1-4$); a, b are geometric isomers. Conditions as in Figure 1. (Reproduced by permission of Elsevier Science Publishers BV)

hydroborates and iodohydroborates (respectively). The zones can be revealed with acridinium hydrochloride or, when radioactive halogens are used, the electropherogram is scanned. Isomers were found to move differently as their form changed from spherical to ellipsoid.

References

[1] R. Guedes de Carvalho, *J. Chromatogr.* **1** (1958) 47.
[2] K. G. Bührens and W. Preetz, *Angew. Chem. Int. Ed. Engl.* **89** (1977) 195.
[3] K. G. Bührens and W. Preetz, *J. Chromatogr.* **139** (1977) 291.

(ii) Condensed phosphates

Condensed phosphates are of biochemical, environmental and industrial interest as well as a fascinating area of inorganic chemistry.

A clear picture of the species which exist in a solution was not obtained until the advent of paper chromatography. Here, furthermore, it was necessary to overcome a technical detail, namely that papers have to be washed free of calcium salts, otherwise phosphates precipitate at the start as insoluble calcium salts.

The original work was done mainly by J. P. Ebel [1], and Figure 4 summarizes the main points of the new discoveries. The homologous series

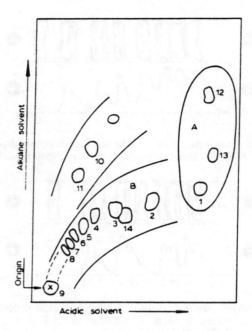

Figure 4. Separation of a mixture of phosphorus oxoacids by two-dimensional paper chromatography. Alkaline solvent, 2-propanol–2-butanol–water–20% ammonia (40:20:39:1); acidic solvent, 2-propanol–water–20% trichloroacetic acid-25% ammonia (700:100:200:3). Compounds: 1 = orthophosphate; 2 = pyrophosphate; 3 = triphosphate; 4 = tetraphosphate; 5 = pentaphosphate; 6 = hexaphosphate; 7 = heptaphosphate; 8 = octaphosphate; 9 = Graham's salt; 10 = trimetaphosphate; 11 = tetrametaphosphate; 12 = hypophosphite; 13 = phosphite; 14 = hypophosphate. (Reproduced from [1] by permission of Consiglio Nazionale delle Ricerche, Ufficio Publicazioni e Informazioni Scientifiche, Rome)

of condensed pentavalent linear polyphosphates, viz. orthophosphate, pyrophosphate, triphosphate etc., separate well in both alkaline and acid solvents. The cyclic phosphates, trimetaphosphate and tetrametaphosphate also separate very well. Different valencies of phosphorus: hypophosphite (valency 1), phosphite (valency 3), phosphate (valency 5) and hypophosphate (valency 4), can also be separated readily without any changes during development. High polymers such as Graham's salt remain at the start. Typical R_F values are shown in Tables 1 and 2.

Figure 5 shows the separation of condensed mixed valency phosphates. This chromatogram was chosen to show that cellulose thin layers also give good separations. The only notable advantage of the thin layers seems to be that of greater speed [5].

Because of the great interest in this field, numerous reviews and several monographs have appeared, e.g. [1, 4, 6].

Table 1. R_F values of polyphosphates in acidic and basic solvents [2]

Paper: Whatman No. 4.
Solvent: Acid solvent: A_1 = Isopropanol, 75 ml; water, 25 ml; trichloroacetic acid, 5 g; ammonia (22°Bè), 0.3 ml.
A_2 = Ethanol, 80 ml; water, 20 ml; trichloroacetic acid, 5 g; ammonia (22°Bè), 0.3 ml.
A = t-Butanol, 70 ml; water, 30 ml; picric acid, 4 g; ammonia (22°Bè), 0.2 ml.
Basic solvent: B_1 = Isopropanol, 40 ml; isobutanol, 20 ml; water, 39 ml; ammonia (22°Bè), 1 ml.
B_2 = Ethanol, 30 ml; n-propanol, 30 ml; water, 39 ml; ammonia (22°Bè), 1 ml.
B_3 = Ethanol, 30 ml; isobutanol, 30 ml; water, 39 ml; ammonia (22°Bè), 1 ml.
Development: Ascending (cylinders).

	Solvent					
	A_1	A_2	A_3	B_1	B_2	B_3
Hypophosphite	0.81	0.84	–	0.72	–	0.75
Phosphite	0.83	0.86	–	0.49	–	0.49
Pyrophosphite	0.83	0.86	–	0.64	–	0.66
Hypophosphate	0.38	0.48	–	0.25	–	0.21
Orthophosphate	0.79	0.80	0.76	0.36	0.45	0.39
Pyrophosphate	0.53	0.62	0.42	0.31	0.36	0.29
Tripolyphosphate	0.34	0.49	0.23	0.29	0.30	–
Trimetaphosphate	0.21	0.33	0.14	0.52	0.74	–
Tetrametaphosphate	0.13	0.16	0.07	0.41	0.61	–
Graham's salt	0	0	0	0	0	–

Table 2. R_F Values of polyphosphates [3]

Paper: Schleicher & Schüll 2040a.
Solvents: S_1 = Isopropanol, 75 ml; water, 25 ml; trichloroacetic acid, 5 g; 20% ammonia, 0.3 ml
S_2 = Isopropanol, 70 ml; water, 10 ml; 20% trichloroacetic acid, 20 ml; 25% ammonia, 0.3 ml.
Development: Ascending.

	Solvent	
	S_1	S_2
Monophosphate $(PO_4)^{3-}$	0.69	0.73
Diphosphate $(P_2O_7)^{4-}$	0.44	0.53
Triphosphate $(P_3O_{10})^{5-}$	0.29	0.39
Tetraphosphate $(P_4O_{13})^{6-}$	0.17	0.29
Pentaphosphate $(P_5O_{16})^{7-}$	0.11	0.22
Hexaphosphate $(P_6O_{19})^{8-}$	0.07	0.16
Heptaphosphate $(P_7O_{22})^{9-}$	0.04	0.11
Octaphosphate $(P_8O_{25})^{10-}$	–	0.08
Trimetaphosphate $(P_3O_9)^3$	0.20	0.32
Tetrametaphosphate $(P_4O_{12})^{4-}$	0.08	0.18

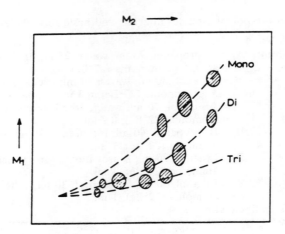

Figure 5. A two-dimensional thin layer separation of condensed phosphates [5]. Schematic representation of actual chromatogram. Mono: $^5P, ^3P, ^1P$; Di: $^4P-^4P, ^5P-O-^5P, ^2P-^4P, ^3P-O-^5P, ^3P-O-^3P$; Tri: $^4P-^3P-^4P, ^3P-O-^4P-^4P, ^3P-O-^5P-O-^5P$. Stationary phase: Cellulose MN300 HR; mobile phases: M_1 = 2-propanol–methanol–TA5–water (7:6:2:5), M_2 = methanol–conc. NH_3–water–trichloroacetic acid (55:5:40:3 g); TA5 = 25 g of trichloroacetic acid dissolved in 100 ml of water of pH 5 (NH_3). Conditions: two-dimensional TLC using M_1 (15 cm in 180 min) and subsequently M_2 (15 cm in 100 min) as mobile phases; plate dried for 5 min in warm air between first and second runs

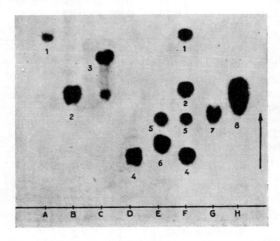

Figure 6. A one-dimensional chromatogram with ethanol–isobutanol–water–conc. NH_3 (30:30:39:1) as solvent [1]. A = hypophosphite (1); B = phosphite (2); C = pyrophosphite (3); D = hypophosphate with a trace of orthophosphate; E = a mixture of orthophosphate (5) and pyrophosphate (6); F = a mixture of hypophosphite (1), phosphite (2), orthophosphate (5) and hypophosphate (4); G = arsenate (7); H = arsenite (8). (Reproduced from [1] by permission of Consiglio Nazionale delle Ricerche, Ufficio Publicazioni e Informazioni Scientifiche, Rome)

We reproduce a one-dimensional paper chromatogram obtained by Ebel (Figure 6), showing that arsenite and arsenate can also be characterized readily in presence of phosphates. Mixed condensed phosphates–arsenates have also been studied by paper chromatography. So too have fluorophosphates and fluoroarsenates [7, 8]) and thiophosphate [9].

Later work shows that condensed phosphates can be separated also on ion exchangers, by HPLC and by electrophoresis. However, little greater

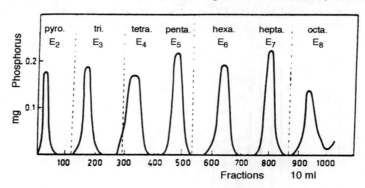

Figure 7. Stepwise elution of polyphosphates from an anion exchange column (Dowex 1-X10, 100–200 mesh) [1]. The eluents used were x M NaCl–0.005 M sodium maleate, pH 6.8; the values of x were 0.18 for E_2, 0.19 for E_3, 0.21 for E_4, 0.23 for E_5, 0.26 for E_6, 0.28 for E_7 and 0.30 for E_8. (Reproduced from [1] by permission of Consiglio Nazionale delle Ricerche, Ufficio Publicazioni e Informazioni Scientifiche, Rome)

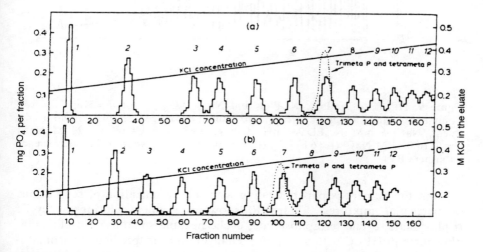

Figure 8. Ion exchange chromatography of a polyphosphate mixture. Ion exchanger, Dowex 1–X4; gradient elution: (a) borate buffer, pH 8.0; (b) ammoniacal buffer, pH 9.3 [10]

insight into the species present and their chemistry has been achieved with the more efficient(?) and certainly more complicated techniques.

A stepwise elution, varying the amounts of NaCl in a pH 6.8 buffer, is shown in Figure 7. A gradient elution separation achieved by Matsuhashi [10] is shown in Figure 8. The latter depicts a better separation of the higher homologues than was obtained on paper chromatograms.

Still better separations of higher homologues (to about 30 units chain length) were obtained by HPLC on an anion exchange resin (Figure 9).

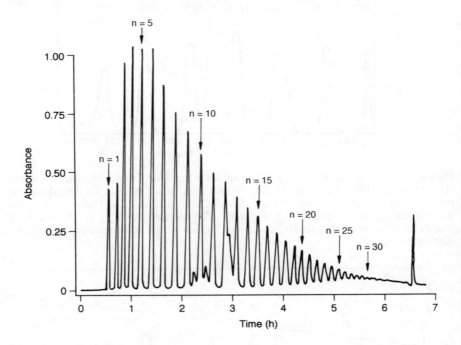

Figure 9. HPLC of Na_3PO_4, $Na_4P_2O_7$ and $Na_{n+2}P_nO_{3n+1}$ (for $\bar{n} = 5$ and $\bar{n} = 10$). Column, 100 mm × 9 mm i.d., Hitachi 2630 anion exchange resin (4% cross-linked); eluent, NaCl–5 mM Na_4EDTA, pH 10.0; gradient, convex from 0.22 M NaCl ($t = 0$) to 0.53 M NaCl ($t = 6$ h) [11]. (Reproduced by permission of Elsevier Science Publishers BV)

From the numerous papers dealing with the optimization of separations of particular mixtures, Figure 10 shows a gradient elution of cyclic triphosphate, cyclic tetraphosphate, cyclic hexaphosphate and cyclic octaphosphate in HPLC with an anion exchanger and the 'predicted' chromatogram obtained by computer calculation which was used to establish the optimum separation. Retention times could be calculated with about 4% error and band width with 5–16% error, Figures 11 and 12 show

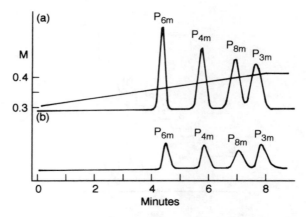

Figure 10. Predicted (a) and observed (b) chromatograms for a mixture of four cyclic polyphosphates. Column, TSKgel SAX (anion exchanger), 500 mm × 4.0 mm i.d.); flow rate: 1.0 ml/min; eluents: A, 0.3 M potassium chloride + 0.1% Na$_4$EDTA (pH 10.2); B, 0.4 M potassium chloride + 0.1% Na$_4$EDTA (pH 10.2); column temperature, 30 °C [12]. (Reproduced by permission of Elsevier Science Publishers BV)

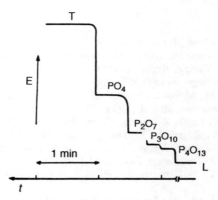

Figure 11. Isotachophoretic separation of linear condensed phosphates having n = 1–4. L = 0.01 M HCl + β-alanine, pH 4.0; T = glutamic acid; E = electrical gradient. (From [13]; reproduced by permission of Elsevier Science Publishers BV)

two isotachophoretic separations, which illustrate that electromigration is an equally good alternative to chromatography for phosphorus oxoacids.

References

[1] J. P. Ebel, in *Metodi di Separazione nella Chimica Inorganica*, Vol. 1, C. N. R., Rome (1963), p. 199.

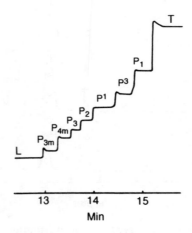

Figure 12. Isotachopherogram of various phosphorus oxoacids: P^1 = phosphinate; P^3 = phosphonate; P_1 = phosphate; P_2 = diphosphate; P_3 = triphosphate; P_{3m} = cyclotriphosphate; P_{4m} = cyclotetraphosphate. L = 0.01 M HCl + histidine, pH 5.5, 0.1% Triton X-100; T = hexanoate. (From [13]; reproduced by permission of Elsevier Science Publishers BV)

[2] J. P. Ebel, *Mikrochim. Acta* (1954) 679.
[3] E. Thilo and H. Grunze, *Die Papierchromatographie der Kondensierten Phosphate*, Akademie Verlag, Berlin, (1955).
[4] K. Gasser, *Mikrochim. Acta* (1957) 594.
[5] M. Baudler and M. Mengel, *Fresenus Z. Anal. Chem.* **211** (1965) 42.
[6] H. Hettler, *Chromatogr. Rev.* **1** (1959) 225.
[7] L. Kolditz and A. Feltz, *Z. Anorg. Chem.* **293** (1957) 155.
[8] L. Kolditz and W. Röhnsch, *Z. Anorg. Chem.* **293** (1957) 168.
[9] E. Steger and U. Seener, *Naturwissenschaften* **46** (1959) 109.
[10] M. Matsuhashi, *J. Biochem. (Tokyo)* **44** (1957) 65.
[11] H. Yamaguchi, T. Nakamuro, Y. Hirai and S. Ohashi, *J. Chromatogr.* **172** (1979) 131.
[12] Y. Baba and G. Kura, *J. Chromatogr.* **550** (1991) 5.
[13] P. Boček and F. Foret, *J. Chromatogr.* **313** (1984) 189.

(iii) Sulphur compounds

POLYTHIONATES

Polythionates, as produced by a reaction between $S_4O_6^{2-}$ and $S_2O_3^{2-}$, have been separated by paper chromatography, paper electrophoresis and thin layer chromatography. A high voltage electropherogram showing a separation of the thionates up to the nonathionate is shown in Figure 13 [1]

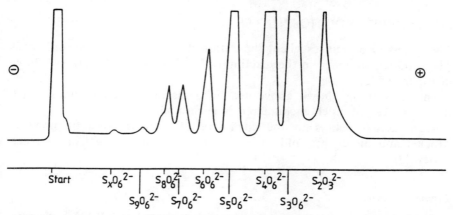

Figure 13. Separation of ^{35}S-labelled polythionates by paper electrophoresis. (Reproduced by permission of Elsevier Science Publishers BV)

SELENOPOLYTHIONATES

These species were separated by high voltage electrophoresis up to the Se$_7$-mer as shown in Figure 14 [1]

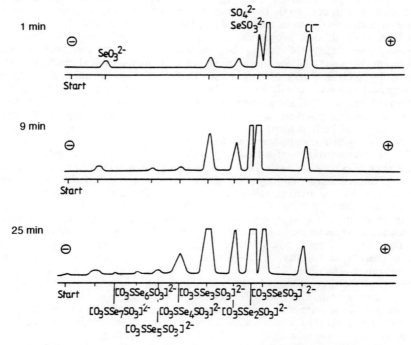

Figure 14. Separation of ^{35}S- and ^{75}Se-labelled selenopolythionates by paper electrophoresis. (Reproduced by permission of Elsevier Science Publishers BV)

POLYSULPHONDIPHOSPHATES

These ions also yield excellent separation in high voltage electrophoresis. The products obtained in the reaction of $^{35}S^{32}PO_3^{3-}$ with $^{34}S_2S_2O_6^{2-}$ are shown in Figure 15 [2].

The separations shown above deal with fascinating compounds which, however, are not much encountered in general chemistry.

More actual results in the chemistry of sulphur were obtained by Möckel and his group, and we would like to quote here from a recent review [3].

Elemental sulphur

The first sulphur ring separated from reaction mixtures by the RPLC technique was the S_8. In these experiments, which were planned for the quantitative analysis of S_8 in aqueous suspensions, two small peaks were observed to elute faster than S_8. Later we found that they represented S_6 and S_7 rings. Further investigations revealed the existence of many hitherto unknown sulphur homocycles, all of which could be separated on ODS phases. RPLC is still the best and sometimes the only way to separate complex mixtures of sulphur rings.

Sulphur rings containing other than eight sulphur atoms are encountered more often than commonly thought; however, their presence is not recognized. They are all referred to as 'elemental sulphur', which, traditionally, is assumed to be S_8. The physical properties, such as molecular and crystalline structure, colour, and density of the various sulphur homocycles are more or less different. Schmidt and Siebert pointed out that their chemical reactivity may differ by several orders of magnitude. If, for example, a reaction produces 'elemental sulphur' consisting mainly of S_6 rings instead of S_8 rings, the chemical behaviour of the solution may be different than expected. Therefore, from a chemical point of view, when referring to 'elemental sulphur' the ring sizes n_s should be indicated.

The thermodynamically stable molecule S_8 is soluble in most organic solvents. The RP chromatogram of a freshly prepared solution shows the solvent peak and that of S_8, which, with ODS phases and pure MeOH eluent, appears at $I_K = 1240 \pm 10$. After standing for a day or more, the solution always contains S_6 and S_7 and sometimes also S_9 to S_{12}. The mechanism of formation is unknown, although some photochemistry is probably involved.

If solutions of S_8 in nonpolar solvents like cyclohexane are ultraviolet-irradiated, larger amounts of all ring sizes from S_6 to about S_{20} are produced.

The sulphur formed when sodium thiosulphate solutions are acidified contains more S_6 and S_7 than S_8. Even medium-sized rings ($n \leq 15$) are produced with appreciable yields.

Sulphur melts contain many different homocycles, some of which have been isolated on a preparative scale for further investigation.

One could mention many more examples of the formation of sulphur rings with $n \neq 8$, several of which are of more than purely academic interest. They are, in fact, so common that it is difficult to obtain pure S_8 solutions!

Typical separations are shown in Figures 16 and 17 on p. 174.

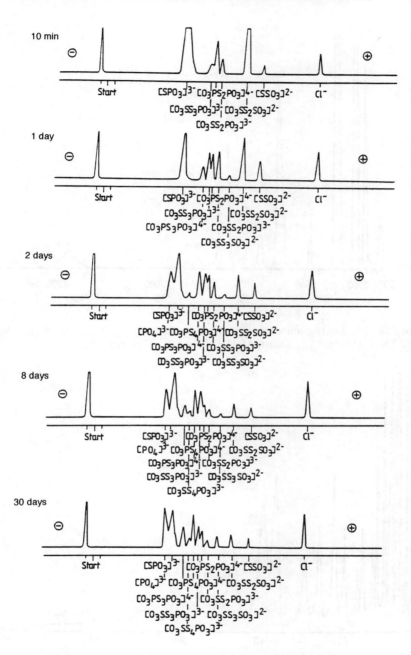

Figure 15. Separation of ^{35}S- and ^{32}P-labelled polysulphondiphosphates and polysulphan phosphene sulphonates by paper electrophoresis. The reaction between ^{35}S^{32}PO$_3^{3-}$ and ^{34}S$_2$S$_2$O$_6^{2-}$ is shown here

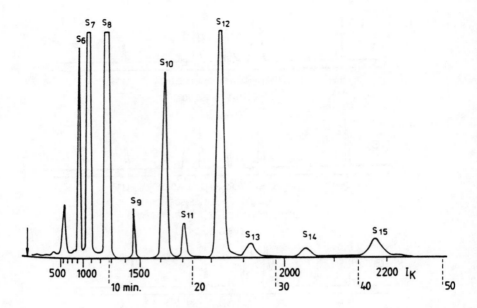

Figure 16. Chromatogram of sulphur homocycles S_6 to S_{15}. Column, 10 cm × 8 mm RadPAK A (10 µm C_{18}); eluent, MeOH; flow rate, 1 ml/min

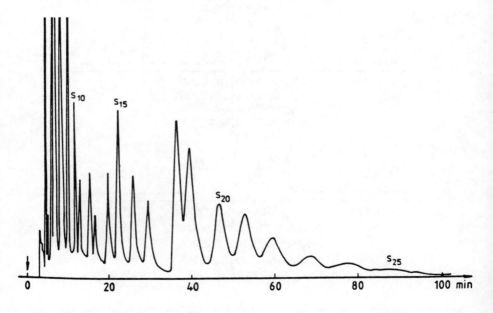

Figure 17. Chromatogram of sulphur homocycles S_6 to S_{25}. Column, 10 cm × 8 mm RadPAK A; eluent, 75% MeOH+25% cyclohexane; flow rate, 1 ml/min

Polysulphanes (H_2S sulphides and polysulphides)

Ammonium polysulphide is a familiar reagent in the analytical laboratory. Attempts to chromatograph it on paper or thin layers has never given satisfactory chromatograms (in my hands). It seems to react with paper and thus 'gets lost' during development.

The situation is very different in Möckel's work with HPLC, from which we quote again.

> The polysulphanes H_2S_n which consist of hydrogen-terminated sulphur chains, are relatively unstable and very sensitive to catalytic activities. In alcoholic (methanol to pentanol) solutions they can be kept for several hours and separated in RP systems. After standing for about 1 day, they have decomposed to yield H_2S and sulphur rings $S_n - 1$. Figure 18 shows the chromatogram of H_2S_2 to H_2S_{14} in methanolic solution. The presence of the sulphur homocycles S_6, S_7, and S_8 can also be seen.

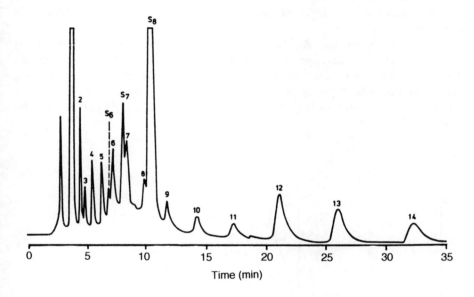

Figure 18. Chromatogram of polysulphanes H_2S_2 to H_2S_{14} (values on peaks give n_S). Column, 10 cm × 8 mm RadPAK A (10 μm C_{18}); eluent, MeOH; flow rate, 1 ml/min

References

[1] E. Blasius, K. Müller and K. Ziegler, *J. Chromatogr.* **313** (1984) 161 and references therein.
[2] E. Blasius and N. Spannhake, *Z. Anorg. Chem.* **399** (1973) 315, 321.
[3] H. J. Möckel, *Ad. Chromatogr.* **26** (1987) 2.

(iv) Halogen acids, especially halogen oxyacids

Halides separate readily in partition as well as ion exchange chromatography, and in most separation systems there is a good separation between chlorate–bromate–iodate.

Among the halogen oxyacids there is one of special interest, namely perbromate. I can strongly recommend reading Prof. Appelman's paper entitled 'Nonexistent compounds: two case histories' [1], from which I quote: 'It is interesting to speculate why certain compounds have resisted discovery for long periods of time............the general pattern seems to be an initial unsuccessful effort to make the compound, followed by detailed rationalization of the failure', and so it happened that a simple inorganic anion, perbromate, was not 'discovered' until 1968!

A short survey of the flat bed methods and spot tests will be given. I realised their relevance only recently when alternative synthesis methods for perbromate were investigated and we had no simple way of detecting the species except a thin layer chromatogram followed by the reaction with methylene blue.

Figure 19 shows the R_F values of halides, halates and perhalates on a strong anion exchange resin paper. Perbromate can be separated from bromate and bromide; it is the strongest adsorbed halogen acid, which is not unexpected, as periodate is not IO_4^- in solution but more likely IO_6^{3-}, and thus perbromate is the largest and least hydrated of the group.

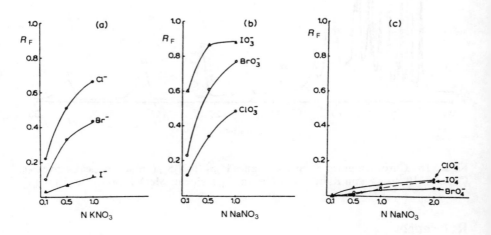

Figure 19. R_F values of (a) halides, (b) halates and (c) perhalates plotted against the $NaNO_3$ concentration (KNO_3 in (a)) on the strongly basic anion exchange resin paper SB-2 (in the nitrate form) [2]. (Reproduced by permission of Elsevier Science Publishers BV)

On cellulose and silica gel thin layers, perbromate is readily separated from perchlorate and periodate, as shown in Tables 3 and 4. The relative speeds in high voltage paper electrophoresis are shown in Table 5. Here another interesting phenomenon can be observed. Iodate migrates cationically in zirconium and thorium salts, and this seems to be due to specific complex formation of iodate with trivalent and tetravalent metal ions. It is also responsible for a very low R_F value for iodate on aluminium oxide thin layers, as shown in Figure 20 (and here compared with the sequence on anion exchangers).

Table 3. R_F Values of halo acids on cellulose thin layers [3]. (Reproduced by permission of Elsevier Science Publishers BV)

Thin layer: Macherey – Nagel Cell 300. Solvent: mixtures of isopropanol and 1.5 N ammonia solution

Ion	Ratio of isopropanol to 1.5 N ammonia solution										
	95:5	90:10	80:20	70:30	60:40	50:50	40:60	30:70	20:80	10:90	5:95
ClO_3^-	0.02	0.34	0.65	0.74	0.84	0.85	0.92	0.90	0.90	0.90	1.00
BrO_3^-	0.05	0.16	0.33	0.29 ↓^{a} 0.54	0.70	0.74	0.84	0.84	0.88	0.88	0.95
IO_3^-	0.07	0.03	0.10	0.00 ↓ 0.30	0.42 ↓ 0.60	0.56	0.00 ↓ 0.70	0.70	0.78	0.90	0.94
ClO_4^-	0.02	0.35	0.62	0.64–0.83	0.86	0.89	0.83	0.80	0.84	0.93	0.90
BrO_4^-	0–0.70	0.67	0.76	0.62 ↓ 0.87	0.75 ↓ 0.91	0.86	0.81	0.84	0.84	0.92	0.91
IO_4^-	0.02	0.00	0.10	0–0.30	0.00 ↓ 0.55	0.00 ↓ 0.60	0.00 ↓ 0.61	0.00 ↓ 0.70	0.00 ↓ 0.79	0.00 ↓ 0.93	0.89

^{a} ↓ = comet.

Table 4. R_F Values of halo acids on silica gel thin layers [3]. (Reproduced by permission of Elsevier Science Publishers BV)

Thin layer: Macherey–Nagel SIL G. Solvent: mixtures of isopropanol and 1.5 N ammonia solution

Ion	Ratio of isopropanol to 1.5 N ammonia solution										
	95:5	90:10	80:20	70:30	60:40	50:50	40:60	30:70	20:80	10:90	5:95
ClO_3^-	0.53	0.57	0.67	0.62	0.64	0.76	0.84	0.85	0.92	0.87	0.85
BrO_3^-	0.44	0.47	0.48	0.61	0.71	0.76	0.76	0.85	0.85	0.90	0.91
IO_3^-	0.04	0.08	0.21	0.33	0.36	0.51	0.66	0.70	0.80	0.86	0.76
ClO_4^-	0.20–0.49	0.62	0.66	0.67	0.76	0.80	0.78	0.88	0.91	0.93	0.93
BrO_4^-	0.71	0.75	0.77	0.76	0.79	0.86	0.88	0.90	0.90	0.92	0.92
IO_4^-	0.02	0.02	0–0.21	0.02	0.03	0.02	0–0.65	0–0.17	0.02–0.16	0–0.90	0–0.73

Table 5. Paper electrophoresis of perhalates and halates [2] (Reproduced by permission of Elsevier Science Publishers BV)

Potential, 1000 V; time, 30 min; temperature, 7–8°C; paper, Whatman No. 1

Salt	Concentration	Electrophoretic movement						Movement relative to ClO_4^-				
		ClO_4^-	ClO_3^-	BrO_4^-	BrO_3^-	IO_4^-	IO_3^-	ClO_3^-	BrO_4^-	BrO_3^-	IO_4^-	IO_3^-
NaOH	0.1 M	−73	−75	−68 T	−64	0	−38	1.03	0.93	0.88	0	0.52
$(NH_4)_2CO_3$	2%	−66	−75	−67	−61	{−36 / −2}	−36	1.14	1.01	0.92	1.39	0.54
NH_4NO_3	0.1 M	−77	−77	−66 T	−65	−38	−36	1.0	0.86	0.84	0.49	0.47
NH_4NO_3	0.5 M	−83	−86	−77	−72	−46	−45	1.04	0.93	0.87	0.55	0.54
$Mg(NO_3)_2$	0.1 M	−82	−79	−71 T	−62	−34	−34	0.96	0.87	0.76	0.41	0.41
$Al(NO_3)_3$	0.1 M	−72	−70	−66 T	−51	−17 T	−17	0.97	0.92	0.71	0.24	0.24
$ZrOCl_2$	0.1 M	−66	−58	−62	−31	{+35–+13 / 0}	+30	0.88	0.94	0.47	–	–
$Th(NO_3)_4$	0.1 M	−73	−68	−66	−47	+29–0	+31	0.93	0.90	0.65	–	–

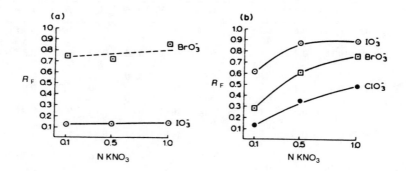

Figure 20. R_F values plotted against the concentration of KNO_3 used as eluent: (a) for alumina thin layers; (b) for SB-2 anion exchange resin paper [4]. (Reproduced by permission of Elsevier Science Publishers BV)

Incidentally, there is no correlation for anions in general between the adsorption on aluminium oxide layers and that on anion exchangers as shown in Figure 21. Ion pairing and complexation on the aluminium oxide seems to play a large role, although several authors assumed that 'ion exchange' with the surface aluminium ions of Al_2O_3 should explain the separations obtained on alumina. For more details see [4].

Spot tests are shown in Table 6. Methylene blue gives a sensitive red spot only with perchlorate and perbromate. Perbromate does not react with KI unless the solution is strongly (6 M!) acid with HCl; it can thus be distinguished readily from all other haloacids by a simple separation plus two spot tests.

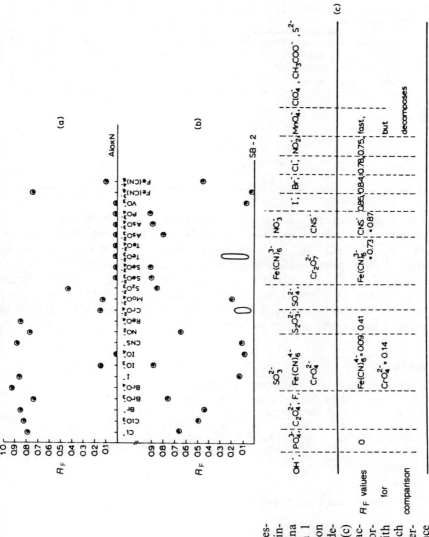

Figure 21. Schematic representation of R_F values of inorganic anions: (a) on alumina thin layers (developed with 1 N NNO_3); (b) on SB-2 anion exchange resin paper (developed with 1 N KNO_3); (c) displacement sequence (according to Kubli [7]) of inorganic anions on alumina with R_F values from (a) below each anion [4]. (Reproduced by permission of Elsevier Science Publishers BV)

Table 6. Spot test reactions of halo acids [3]. (Reproduced by permission of Elsevier Science Publishers BV)

Ion	Manganous sulphate + phosphoric acid	Manganous sulphate + sulphuric acid	Acidified KCNS	Acidic pyrogallol	Potassium iodide in HCl	Methylene blue
BrO_4^-	+	–	–	–	+	+
BrO_3^-	+	+	–	+	+	–
IO_4^-	+	–	–	–	+	–
IO_3^-	–	–	+	+	+	–
ClO_4^-	–	–	–	–	–	+
ClO_3^-	+	–	–	–	+ (in 6 N HCl)	–

In the work of Appelman [1, 5, 6], perbromate could be formed only by oxidation with fluorine or xenon fluoride and in a hot atom process: $^{83}SeO_4^{2-} \rightarrow {}^{83}BrO_4^- + \beta^-$. All the usual oxidation reactions were unsuccessful, except for some poor yields in electrochemical oxidations. This, however, does not rule out the formation of some perbromate by natural processes such as strong UV irradiation or cosmic radiation. Perbromate is very stable, even in acid solutions, and I would not be surprised if it were found in nature if one really looked for it.

References

[1] E. H. Appelman, *Acc. Chem. Res.* **6** (1973) 113.
[2] M. Lederer and M. Sinibaldi, *J. Chromatogr.* **60** (1971) 275.
[3] L. Ossicini and M. Balzoni, *J. Chromatogr.* **79** (1973) 311.
[4] M. Lederer and C. Polcaro, *J. Chromatogr.* **84** (1973) 379.
[5] E. H. Appelman, *J. Am. Chem. Soc.* **90** (1968) 1900.
[6] E. H. Appelman, *Inorg. Chem.* **8** (1969) 223.
[7] H. Kubli, *Helv. Chim. Acta* **30** (1947) 453.

(v) Rhenium and technetium

RHENIUM

Potassium perrhenate $K^+ReO_4^-$ is quite stable in aqueous solutions and is not reduced by HCl or HBr. Careful reduction with $SnCl_2$ solution can yield an unstable Re(V) and a more stable $Re(IV)Cl_6^{2-}$. The three were separated by Pavlova [1] by anion exchange chromatography on paper (Figure 22).

$(NH_4)_2 Re(IV)Cl_6$ is quite stable as a solid; in 1 N HCl $Re(IV)Cl_6^{2-}$ hydrolyses slowly at room temperature to yield a mixture of aquochlorocomplexes. These are separated well both by paper electrophoresis and on anion exchange cellulose papers [2]; see Figures 23–25. While paper electro-

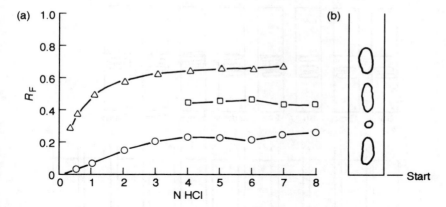

Figure 22. (a) R_F–HCl concentration diagram for various valencies of rhenium on Whatman DE-20 paper developed at room temperature. ReO_4^-(△——△); Re(V) (□——□); $ReCl_6^{2-}$ (○——○). (Reproduced by permission of Elsevier Science Publishers BV) (b) Separation of Re(VII), Re(V), Sn(IV) and Re(IV) (top to bottom) on Whatman DE-20 paper developed with 4 N HCl at 0 °C. (Reproduced by permission of Elsevier Science Publishers BV)

phoresis produces two anionic and one neutral species, there are more spots on the ion exchange paper which could possibly be *cis–trans* isomers. This work seems to merit further investigation.

When dissolved in HBr, the hexachlororhenate yields mixed chlorobromorhenates (IV), which are separated well on cellulose adsorption thin layers and by HPLC (see pages 45 and 125).

TECHNETIUM

Technetium does not occur in nature and has no inactive isotope: ^{99}Tc (half life 2.1×10^5 years) is the usual isotope encountered and is prepared from fission products (according to Heslop and Robinson [3]) as follows.

$$\text{Fission products} \xrightarrow{\text{HCl}} \begin{cases} UCl_4 \\ + \\ \text{fission product chlorides} \end{cases} \xrightarrow{H_2O_2} \begin{cases} UO_2Cl_2 \\ + \\ \text{Chlorides} \end{cases}$$

add $PtCl_4$ ↓ then pass H_2S

$$Tc_2O_7 \xleftarrow[18\ NH_2SO_4]{\text{Distil with}} \text{Residue} \xleftarrow[\substack{\text{ammoniacal } H_2O_2 \\ \text{and evaporate to} \\ \text{dryness with } Br_2}]{\text{Dissolve } Tc_2S_7} \begin{cases} PtS_2, Tc_2S_7 + \text{other} \\ \text{acid–insoluble sulphides} \end{cases}$$

Ammonium pertechnetate is made on the macro scale. A pile operating at 100 MW yields about 2.5 g of ^{99}Tc per day, so the annual world

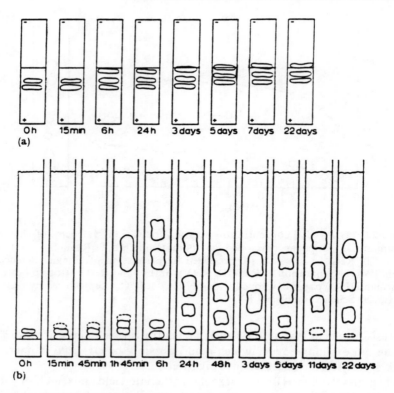

Figure 23. Ageing of a solution of $(NH_4)ReCl_6$ in 1 N HCl. (a) Paper electrophoresis on Whatman No. 1 paper (20 cm × 4 cm) by the glass plate technique at 200 V for 30 min with 0.3 N HCl as electrolyte; (b) ion exchange cellulose chromatography, Macherey–Nagel strong anion exchange cellulose paper developed with 0.5 N HCl. Reagent: $SnCl_2$–KCNS in ≈ 3 N HCl. (Reproduced by permission of Elsevier Science Publishers BV)

production amounts to thousands of kilograms. Technetium-99 is a weak beta emitter and a millicurie (about 20 mg of $KTcO_4$) is about as dangerous a radiochemical as 100 grams of uranyl acetate.

Pertechnetate is reduced readily in conc. HCl or conc. HBr, first to Tc(V) and then to $Tc(IV)X_6^{2-}$.

Figure 26, from the work of Shukla [4], shows the adsorption paper chromatography of Tc solutions in HCl as well as the elegant separation of all three valencies. Tc(V) was shown to be quite stable at low temperatures and is not completely reduced in HCl even after heating in conc. HCl for 4 hours.

In HBr reduction to Tc(IV) proceeds more readily, as shown in Figure 27. Only in 20% and 40% HBr is TcO_4^- not reduced at room temperature [5].

Like $ReCl_6^{2-}$, $TcCl_6^{2-}$ is hydrolysed readily in dilute HCl to yield a mixture of chloroaquo species and finally to form insoluble TcO_2.

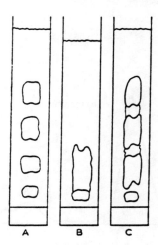

Figure 24. Comparison of various anion exchange papers for the separation of aquochlororhenates(IV). A, Macherey–Nagel strong anion exchange cellulose paper; B, Whatman AE-30 (aminoethylcellulose) paper; C, Whatman DE-20 (diethylaminoethylcellulose) paper. All papers were developed with 0.5 N HCl. Reagent as in Figure 23. (Reproduced by permission of Elsevier Science Publishers BV)

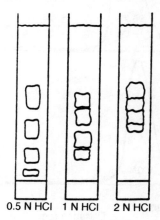

Figure 25. Separation of aquochlororhenates with various concentrations of HCl as eluent. Paper: Macherey–Nagel strong anion exchange cellulose. From left to right 0.5 N, 1 N and 2 N HCl. Reagent as in Figure 23. (Reproduced by permission of Elsevier Science Publishers BV)

Trivalent technetium seems to be formed when technetium (VII), (V) or (IV) reacts with–SH groups in thiourea, thioglycollic acid or cysteine. Orange–yellow complexes are formed, which are adsorbed on cellulose and are sufficiently coloured to serve as spot tests. See Jasim *et al.* [6], Beckmann and M. Lederer [7] and Morpurgo [8].

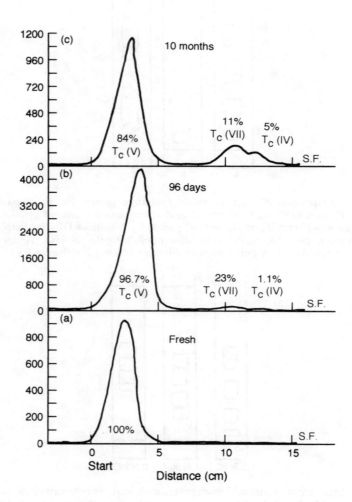

Figure 26. Radiochromatograms on HCl-washed Whatman 3MM (pure cellulose) paper at 8 °C in 0.6 N HCl. (a) Fresh solution of ammonium pertechnetate in conc. HCl at 0 °C; (b) solution (a) aged for 96 days at 0 °C; (c) solution (a) aged for 10 months at 0 °C; (d) solution (c) heated on water bath for 5 minutes; (e) solution (c) heated on water bath for 10 minutes; (f) solution (c) heated on water bath for 20 minutes; (g) solution (c) heated on water bath for 4 hours

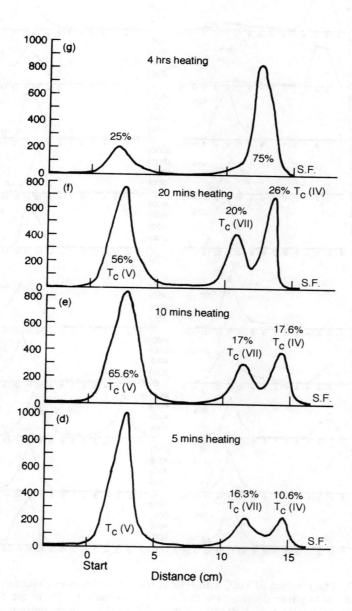

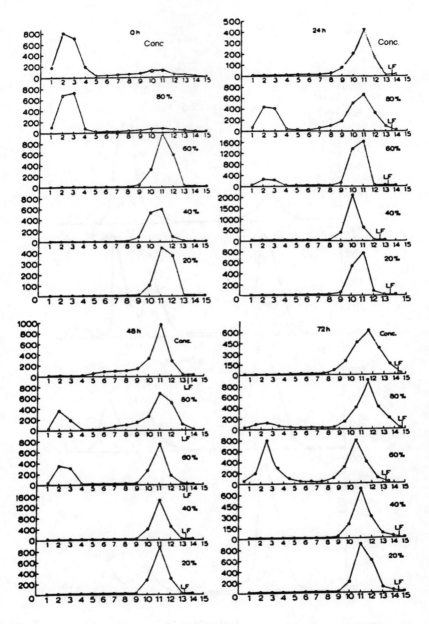

Figure 27. The reaction of TcO_4^- with HBr at room temperature. Chromatograms on Whatman 3MM paper developed with 0.9 N HBr as solvent. From top to bottom: conc. HBr, 80% HBr, 60% HBr, 40% HBr and 20% HBr. On cellulose paper, Tc(V) is separated from all Tc(VII) species. LF = liquid front. (Reproduced by permission of Elsevier Science Publishers BV)

Short-lived (half life 6 hours) ^{99m}Tc is usually prepared via a 'cow'; i.e. it is eluted from ^{99}Mo (as molybdate) adsorbed on a column of Al_2O_3 simply by washing with water. It is used in radiopharmacy for thyroid scintography. According to Shukla [9] traces of aluminium in the eluate can alter the absorption of TcO_4^- in the thyroid. He recommends control of the eluate from the cow by paper chromatography for each diagnostic preparation [9].

References

[1] M. Pavlova, *J. Chromatogr.* **51** (1970) 346.
[2] G. Bagliano and L. Ossicini, *J. Chromatogr.* **19** (1965) 423.
[3] R. B. Heslop and P. L. Robinson, *Inorganic Chemistry* Elsevier, Amsterdam (1960), p. 455.
[4] S. K. Shukla, *Ric. Sci.* **36** (1966) 725.
[5] L. Ossicini, F. Saracino and M. Lederer, *J. Chromatogr.* **16** (1964) 524.
[6] F. Jasim, R. J. Magee and C. L. Wilson, *Talanta* **2** (1959) 93.
[7] T. J. Beckmann and M. Lederer, *J. Chromatogr.* **5** (1961) 341.
[8] L. Morpurgo, *Inorg. Chim. Acta* **2** (1968) 169.
[9] S. K. Shukla, *Quad. Cromatogr.* **1** (1990) 117.

(vi) Ruthenium

Ruthenium is one of the platinum metals which forms stable complexes with valencies from 0 to 8. The octavalent oxide RuO_4 (like OsO_4) is volatile. In aqueous acid solutions the common valencies are 3 and 4. Interest in the solution chemistry of ruthenium is mainly due to ^{106}Ru (a beta-emitter, half life one year), which is one of the major fission products and perhaps the most problematic.

SOLUTION CHEMISTRY IN HCl

A solution of $[Ru(III)H_2OCl_5]^{2-}$ in 0.1 N HCl hydrolyses to give a mixture of anionic, neutral and cationic species, very much like $[RhCl_6]^{2-}$ (see page 194). Shukla [1] studied such a solution by recording the absorption spectrum and electropherograms over 10 days (see Figure 28).

SOLUTION CHEMISTRY IN HNO_3

This has been studied by numerous workers, as spent uranium fuel is treated with HNO_3 to recover the uranium by solvent extraction, leaving the fission products in HNO_3 solution.

Here ruthenium is remarkable by forming very stable mononitroso complexes RuNO, which can then coordinate with nitro, nitrato, aquo and hydroxo groups on the remaining five coordination positions.

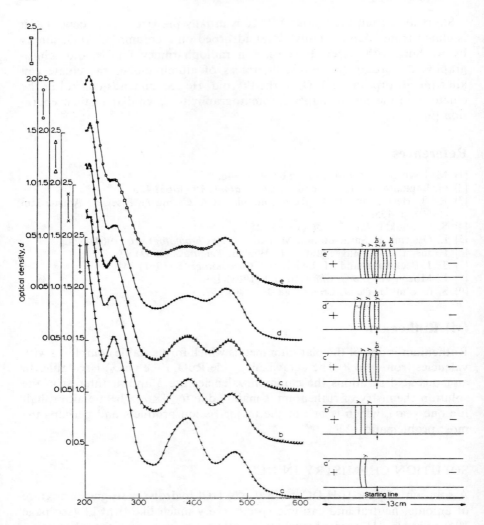

Figure 28. Effect of ageing at room temperature on the absorption curve (a–e) and on the electropherogram (a′–e′) of a solution of $[Ru(H_2O)Cl_5]^{2-}$ in 0.1 N hydrochloric acid. Abbreviations used for denoting the colour of the bands: y = yellow; ybr = yellowish brown; br = brown; rbr = raspberry red. a, a′: fresh solution; b, b′: solution aged for 24 h; c, c′: solution aged for 3 days; d, d′: solution aged for 6 days; e, e′: solution aged for 10 days [1]. (Reproduced by permission of Elsevier Science Publishers BV)

A solution of nitrosylruthenium refluxed in 3–4 N HNO_3, after standing for one week, yields an electropherogram as shown in Figure 29, usually with six well separated cationic species (Morpurgo [2]).

Figure 29. Paper electropherogram of a solution of ruthenium nitrosyl complexes present in 3 N HNO$_3$ solution; 1100 V for 40 min with 0.1 N HNO$_3$–0.5 N NaNO$_3$ as electrolyte [2]

In thin layer chromatography on cellulose (MN-Polygram Cel 300) as well as in paper chromatography with butanol–3 N HNO$_3$, there are four or five distinct zones, as shown in Figure 30 [3].

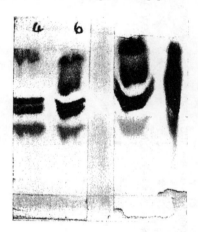

Figure 30. Thin layer chromatograms of ruthenium nitrosyl complexes in 4 N, 6 N, 8 N and 14 N HNO$_3$. Thin layer MN-Polygram Cel 300; solvent: n-butanol–3 N HNO$_3$ (1:1). The zones are detected by holding the layer over a 20% solution of ammonium sulphide [3]

In addition to monomers, polymeric species can form in dilute nitric acid solution. They have been separated by gel filtration (see page 81).

Blasius *et al.* [4] set themselves the formidable task of separating all the monomeric ruthenium nitrosyl nitratoaquo (or hydroxo) complexes; see Figure 31. In addition to high voltage paper electrophoresis, they used isotachophoresis with various amines as spacers (Figure 32) and continuous isotachophoresis.

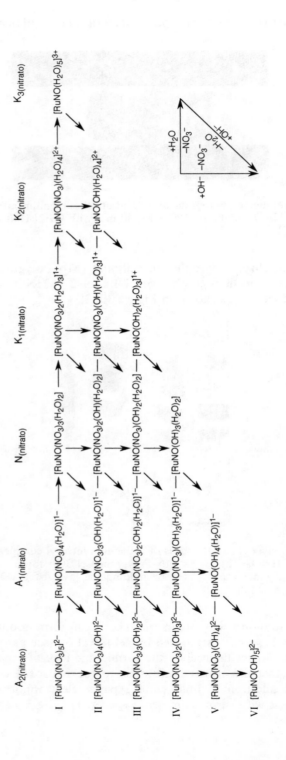

Figure 31. Compilation of the monomeric ruthenium nitrosylnitrato complexes without consideration of the possible positional isomers [4]. (Reproduced by permission of Elsevier Science Publishers BV)

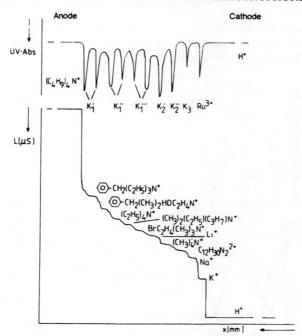

Figure 32. Separation of cationic ruthenium nitrosylnitrato complexes from an aged solution. Capillary isotachophoresis with addition of the spacers shown; 11 000 V for 50 min at 5 °C in a capillary 610 mm long [4]. The explanation of K_1' etc. is shown in Figure 33. (Reproduced by permission of Elsevier Science Publishers BV)

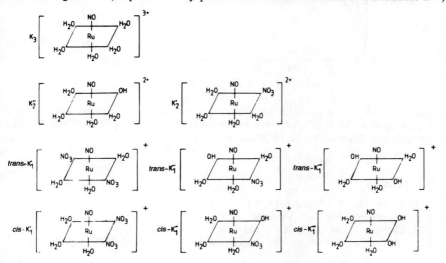

Figure 33. The abbreviations used in Figures 32 and 34 for the positional isomers of the cationic ruthenium nitrosylnitrato complexes [4]. (Reproduced by permission of Elsevier Science Publishers BV)

The evolution of the various complexes (Figure 33) present in a 0.01 M HNO_3 solution at room temperature over 70 days is shown in Figure 34. Further results are given in Figure 35.

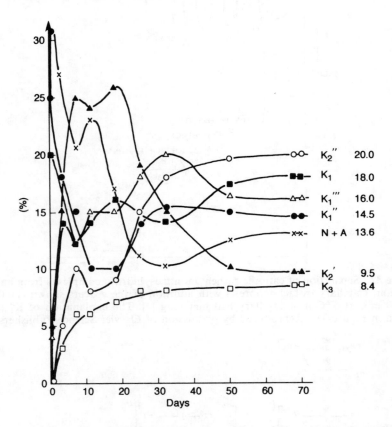

Figure 34. The evolution of complexes present in a ruthenium nitrosylnitrato solution in 0.01 M HNO_3 at room temperature over 70 days [4]. (Reproduced by permission of Elsevier Science Publishers BV)

References

[1] S. K. Shukla, *J. Chromatogr.* **8** (1962) 96
[2] L. Morpurgo, *Ric. Sci.* **38** (1968) 459.
[3] L. Morpurgo, *Ric. Sci.* **36** (1966) 553.
[4] E. Blasius, K. Müller and K. Ziegler, *J. Chromatogr.* **313** (1984) 161.

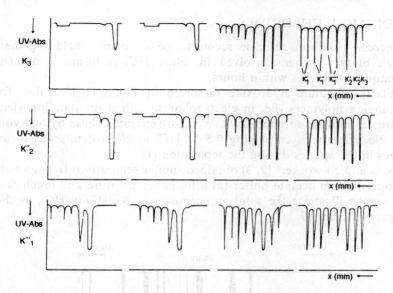

Figure 35. A study of the equilibria reached by single ruthenium nitrosylnitrato complexes under different conditions. Capillary isotachophoretic separations (10 000 V for 40 min at 5 °C in a capillary 610 mm long) are shown. From left to right: 3 h at 0 °C, 45 days at −36 °C, 45 days at 3 °C and 4 h at 100 °C. The symbols used for the complexes are given in Figure 33 [4]. (Reproduced by permission of Elsevier Science Publishers BV)

(vii) Rhodium

Metallic rhodium and rhodium-plated objects are not attacked at all by acids or aqua regia, even with boiling for hours. In order to produce a soluble form of the metal, it must be treated with fused $KHSO_4$ at elevated temperature in a porcelain crucible heated with a Meker burner, or heated in a quartz tube to red heat in a stream of chlorine gas. My involvement in Rh(III) chemistry started in 1951 when I tried to separate carrier-free ^{102}Rh from irradiated ruthenium. We never achieved this separation by a chromatographic or electrophoretic technique, as both these metals form a multitude of stable complexes in solution.

The stable valency of rhodium in solution is Rh(III), which is very similar to Cr(III) in its complexing behaviour. For the chromatographer it has a very interesting and sensitive colour reaction with stannous chloride in KI–HCl, which on heating yields purple spots with most complexed forms of Rh(III). Thus the solution chemistry of Rh(III) has been extensively investigated by paper electrophoresis, ion exchange chromatography and preparative electrophoresis; Cr(III), which lacks a convenient spot test, has been much less widely studied.

RHODIUM(III) CHLORIDES

Commercial rhodium chloride seems to be a polymer held together by halogen bridges. When dissolved in dilute HCl it forms a mixture of chloroaquo complexes within hours.

Similarly, rhodium hydroxide dissolved in dilute HCl at first forms slow-moving polymers (i.e. in electrophoresis), then on standing yields a mixture of chloroaquo complexes. We first separated these by low voltage paper electrophoresis employing 0.5 M HCl as electrolyte so as to avoid changes in the species during the separation [1].

Blasius and co-workers [2, 3] obtained similar separations by high voltage electrophoresis in acetate buffer (at a lower temperature and much faster). Their results illustrate the solution chemistry of Rh(III) in HCl, as shown in Figures 36 and 37.

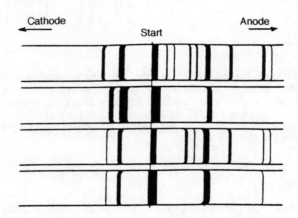

Figure 36. Paper electropherograms of solutions of rhodium chloride in water or in 0.1 M HCl. Paper: Schleicher & Schüll 2043 b Mgl; electrolyte 0.3 M acetic acid–0.2 M sodium acetate; 37.5 V/cm applied for 30 min at 0 °C. The spots were detected with stannous chloride–KI reagent [3]. From top to bottom: a fresh solution in water, an aged solution in water, a fresh solution in 0.1 M HCl, an aged solution in 0.1 M HCl. (Reproduced by permission of Elsevier Science Publishers BV)

The purple hexachlororhodate, which is stable as a solid, seems to hydrolyse quickly once dissolved in aqueous HCl. Depending on the HCl concentration, there is a different quantitative distribution of the chloroaquo species, but in all solutions there are three or four species.

RHODIUM(III) PERCHLORATE

C. K. Jørgensen described a method for preparing the hexaquorhodium (III) ion by boiling rhodium chloride with a large excess of perchloric acid

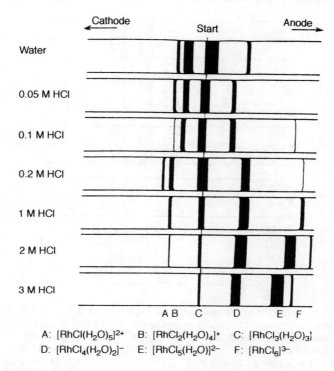

Figure 37. Paper electropherograms of solutions of Rh(III) aged in the concentration of HCl indicated above. Conditions as in Figure 36 [3]. (Reproduced by permission of Elsevier Science Publishers BV)

[4]. A brownish yellow solution is obtained which, however, separates into two species by paper electrophoresis with 1 M $HClO_4$ as electrolyte.

In attempting to prepare a 'pure' hexaquorhodium(III) solution we tried to dissolve rhodium hydroxide in $HClO_4$. The rhodium hydroxide was made by adding an excess of NaOH to a solution of rhodium chloride. Precipitation does not occur immediately, but the mixture changes colour from red through orange to yellow and after about an hour forms a yellow precipitate.

The solution in $HClO_4$ yields three well separated bands, which could be ascribed to (probably) the dichloro-, the monochloro- and the hexaquorhodium(III). Thus a rather interesting phenomenon takes place. When 'rhodium hydroxide' precipitates some of the chloro-groups remain attached to the Rh(III) and on redissolution are still attached.

After four reprecipitations and redissolutions in $HClO_4$, a solution is obtained which has one single fast-moving Rh(III) species (Figure 38). Shukla measured the absorption spectrum of this solution and compared it with the results of Jørgensen [4] and of Ayres and Forrester [5]. However

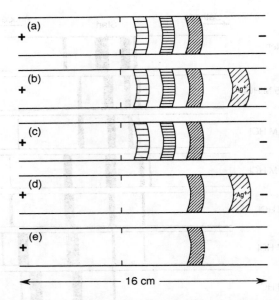

Figure 38. Paper electropherograms in 1 M $HClO_4$ with 250 V for 45 min (Shukla [8]). From top to bottom: (a) rhodium hydroxide (obtained from rhodium chloride and NaOH solution) dissolved in 1 M $HClO_4$; (b) to this solution silver perchlorate was added (without heating); (c) solution (a) was heated for 45 min at 100 °C; (d) solution (b) was heated for 10 min at 100 °C; (e) solution (a) was reprecipitated four times with NaOH and the precipitate was redissolved in 1 M $HClO_4$

in retrospect, I wonder, whether any of the spectra were really that of the hexaquo ion.

It could be possible that in all cases only a monohydroxypentaquorhodium(III) ion was obtained and that a hexaquorhodium species is too unstable to exist in solution, just as hexachlororhodium(III) hydrolyses within minutes to form mainly monoaquopentachlororhodate, even in 6 N HCl.

The analysis of isolated $Rh(H_2O)_6 (ClO_4)_3$ by Shukla seems to indicate that the hexaquo compound was effectively made [6]; however, the whole subject would merit reinvestigation.

POTASSIUM TRISOXALATORHODATE

According to Delepine [7], this compound is obtained by boiling $K_2[RhCl_5 H_2O]^{2-}$ or $K_3[RhCl_6]^{3-}$ for two hours with potassium oxalate in a mole ratio of 1:3. A solution so prepared yields three well separated bands in electrophoresis with 1 M potassium oxalate as electrolyte. Figure 39 shows that another four hours of boiling are required to produce a complete conversion to the $[Rhox_3]^{3-}$ ion.

ELEMENTS AND THEIR CHROMATOGRAPHY AND ELECTROPHORESIS 197

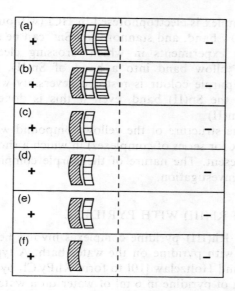

Figure 39. Electropherograms of the reaction of rhodium chloride with potassium oxalate. Electrolyte 1 M potassium oxalate; 250 V applied for 45 min [6]. From top to bottom: (a) a solution heated for 15 min; (b) same solution after heating for 1 h 15 min; (c) after 2 h 15 min; (d) after 4 h; (e) after 5 h; (f) after 6 h

RHODIUM(III) SULPHATE COMPLEXES [8]

The yellow species presumed to be hexaquorhodium(III) migrates anionically in sulphate as electrolyte, as do also $Cr(H_2O)_6^{3+}$, $Al(H_2O)_6^{3+}$ etc, which all form strong outer sphere complexes with sulphate via hydrogen bonds.

On heating solid yellow rhodium sulphate, water is lost at 110 °C and at 190 °C. Electropherograms of this 'heated' green form in acid sulphate are rather disappointing: several zones and long tails are formed. Presumably there are relatively fast equilibria between monodentate and bidentate sulphate as ligand. The electropherograms were all run in sulphate at low pH to prevent hydrolysis. There seems still to be room for further investigation of the sulphate complexes of Rh(III).

COMPLEXES OF Rh(III) WITH STANNOUS CHLORIDE IN HCl

The reaction of stannous chloride in HCl–KI, or HBr–HCl yields highly coloured solutions; this also provides convenient spot tests in electrophoresis and chromatography.

When the (usually) purple solution is electrophorized in a mixture of $SnCl_2$–HCl as electrolyte, the red complex moves anionically [6]. In previous work, G. H. Ayres [9] had proposed a cationic structure.

If the purple complex is electrophoresed in HCl (without $SnCl_2$), it yields an intensively yellow band, and stannous ion Sn^{2+} can be shown to migrate away from it. In experiments in which 'crossing electrophoresis', i.e. migration of the yellow band into a band of Sn(II), is arranged, one observes that the purple colour is restored reversibly while the rhodium complex traverses the Sn(II) band, even if this is done repeatedly with several bands of Sn(II).

Subsequently, the structure of the yellow compound was shown to be a Rh–Sn(II) complex (or series of complexes) in which a rhodium–tin (metal–metal) bond is present. The nature of the purple complex with yet more Sn(II) still awaits investigation.

COMPLEXES OF Rh(III) WITH PYRIDINE

Most syntheses of Rh(III)–pyridine complexes involve heating mixtures of rhodium chloride with pyridine on the water bath. A typical attempt was that by Collman and Holtzclaw [10] to form $RhPy_3Cl_3$ by heating 0.05 g of '$RhCl_3$' and 1.14 g of pyridine in 6 ml of water on a water bath.

The electropherogram in Figure 40 shows that this synthesis yields a mixture of two cationic bands and a neutral and an anionic band [6]. More recent work (unpublished) showed that this solution yields six well separated spots on a paper chromatogram developed with butanol–3 N HCl.

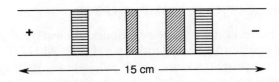

Figure 40. Paper electropherogram of the solution obtained in the synthesis of rhodium–trichlorotripyridine. Electrolyte 1 M $HClO_4$ for 45 min at 250 V [6]

References

[1] M. Lederer, *Zh. Neorg. Khim.* **3** (1958) 1799.
[2] E. Blasius and W. Preetz, *Chromatogr. Rev.* **6** (1964) 191.
[3] E. Blasius, K. Müller and K. Ziegler, *J. Chromatogr.* **313** (1984) 161.
[4] C. K. Jørgensen, *Acta Chem. Scand.* **10** (1956) 500.
[5] G. H. Ayres and J. S. Forrester, *J. Inorg. Nucl. Chem.* **3** (1957) 365; *J. Phys. Chem.* **63** (1959) 1979.
[6] S. K. Shukla, Doctoral Thesis, Paris (1962).
[7] M. Delepine, *Bull. Soc. Chim. Fr.* **29** (1921) 656; *An. R. Soc. Esp. Fis. Quim.* **27** (1929) 485.
[8] S. K. Shukla and M. Lederer, *J. Less-Common Met.* **1** (1959) 255.
[9] G. H. Ayres, *Anal. Chem.* **25** (1953) 1622.
[10] J. P. Collman and H. F. Holtzclaw, Jr., *J. Am. Chem. Soc.* **80** (1958) 2054.

(viii) The rare earths

The rare earths, together with the very similar Group 3A elements Sc, Y and Ac, total 18 elements (almost 20% of our natural 92 elements). It is one of the major achievements of chromatographic methods that these 18 elements can be separated analytically as well as preparatively on an industrial scale.

Before the 1940s the rare earths were mainly separated by fractional crystallization, usually as sulphates or double sulphates. Table 7 shows some solubilities of rare earth salts, which illustrate that complete separation can hardly be possible. Some rare earth elements could be separated in their stable higher or lower valency (than the usual trivalent), notably Ce(IV) and Eu(II), but for the rest most preparations were hopelessly contaminated by the neighbouring element(s). I remember working with an old sample of didymium (didymium was a name for praseodymium–neodymium), which on electrophoresis was revealed as containing about one third of lanthanum.

Table 7. Solubilities of typical salts in water

Cation	Solubility, g/100g H_2O		
	$Ln_2(SO_4)_3 \cdot 8H_2O$ (20°C)	$Ln(BrO_3)_3 \cdot 9H_2O$ (25°C)	$LnCl_3 \cdot 6H_2O$ (20°C)
Y^{3+}	9.76	–	217.0
La^{3+}	–	462.1	–
Ce^{3+}	9.43[a]	–	–
Pr^{3+}	12.74	196.1	–
Nd^{3+}	7.00	151.3	243.0
Pm^{3+}	–	–	–
Sm^{3+}	2.67	117.3	218.4
Eu^{3+}	2.56	–	–
Gd^{3+}	2.89	110.5	–
Tb^{3+}	3.56	133.2	–
Dy^{3+}	5.07	–	–
Ho^{3+}	8.18	–	–
Er^{3+}	16.00	–	–
Tm^{3+}	–	–	–
Yb^{3+}	34.78	–	–
Lu^{3+}	47.27	–	–

[a] Calculated as anhydrous salt.

If we look at the ionic radii of the series the difficulty becomes clear: the differences between Sc–Y–La (radii 0.68, 0.88 and 1.061 A° respectively) are 0.2 and 0.181. A similar series in the Periodic Table would be Na–K–Rb or Ca–Sr–Ba. Separation in all three groups is possible with both partition systems and ion exchange systems. However, the difference between the radii of the rare earths (Figure 41) is only about 0.15, viz. one tenth of that of the Sc – Y or Y – La pairs.

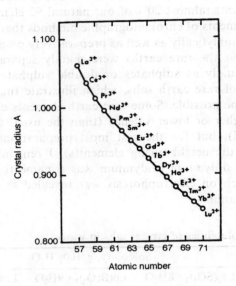

Figure 41. Crystal radii of Ln^{3+} ions

The breakthrough to the complete separation of all the 18 elements came during the Second World War and, being 'classified information', was not published until it was released in 1947. A collection of papers then appeared in Volume 69 of the *Journal of the American Chemical Society*. It contains not only the basic principles of combining ion exchange with complex formation but also much of our present theory of elution and displacement chromatography, as well as the chemical evidence for the identification of element 61, now called promethium, one of the lighter rare earths of which no stable isotope exists in nature.

The optimum separation conditions were established by batch experiments to find the best separation factors between pairs of adjacent rare earths. The variables studied were the kind of complexant, the pH i.e. the ionization of the complexant and/or its concentration, the temperature etc.

The first successful separation with radioactive rare earths is shown in Figure 42. A sulphonic resin, citrate at a controlled pH and elution at 100 °C were used. The high temperature was considered necessary so as to achieve fast equilibrium between the resin and the solution. This required a 'degasser' column to rid the eluent of air bubbles. The effluent was scanned automatically, i.e. continuously, with an end-window counter.

The set-up was thus probably the first 'automatic' chromatograph to be assembled.

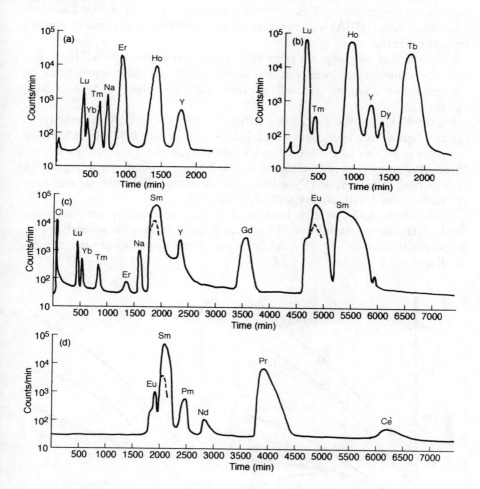

Figure 42. Elution curves of rare separations effected with a 270–325 mesh Dowex-50 column at 100 °C. Bed dimensions 97 cm × 0.26 cm^2; flow rate 1.0 ml/cm^2/min, except in A where 2.0 ml/cm^2/min was used (Ketelle and Boyd [1]). (a) Fractionation of activities produced by neutron irradiation of 0.8 mg of spectrographic grade Er$_2$O$_3$ (Hilger) (pH 3.20); (b) fractionation of heavy rare earth mixture consisting of 0.1 mg each of Lu$_2$O$_3$, Yb$_2$O$_3$, Ho$_2$O$_3$ and Tb$_2$O$_3$ (Tm, Er, Y and Dy present as impurities, pH 3.20); (c) fractionation of intermediate rare earth mixture consisting of 0.1 mg of Ho$_2$O$_3$ and 1.0 mg each of Dy$_2$O$_3$, Gd$_2$O$_3$, Eu$_2$O$_3$ and Sm$_2$O$_3$ (Cl, Lu, Yb, Tm, Er and Na present as impurities, pH 3.25 for 4550 min, then pH 3.33); (d) fractionation of light rare earth mixture consisting of 0.1 mg each of Sm$_2$O$_3$ and Nd$_2$O$_3$ plus 0.01 mg each of Pr$_2$O$_3$, Ce$_2$O$_3$ and La$_2$O$_3$ (Eu present as impurity, Pm produced by 1.7 h ^{149}Nd → ^{149}Pm; pH 3.33 for 1610 min, then pH 3.40)

In later work the separation factors of citrate and other organic acids were compared. Initially lactic acid was considered the best and was also used in electrophoresis. Later, α-hydroxyisobutyric (2-hydroxy-2-methylpropionic) acid (HIBA) was considered, and seems to be the best in various separation systems.

However, not only organic acids yield good separation factors. Nitric acid in a non-aqueous medium, e.g. methanol, yields excellent separations on ion exchange columns as well as ion exchange paper and in solvent extraction.

Figure 43 shows the partition coefficients between tributyl phosphate (a typical liquid ion exchanger) and nitric acid. With 0.98 M HNO_3 the sequence is similar to the sequence of 'non-complexed' rare earths in paper chromatography (e.g. with ethanol – 10% 2 N HCl) as shown in Figure 44. In high concentrations of HNO_3 the sequence is that of the atomic numbers, as also obtained with citrate and lactic acid etc.

Nitriloacetic acid has also been used as a complexant in ion exchange. Table 8 shows the stabilities of rare earth complexes with several aminepolycarboxylic acids. There is an increase of stability with atomic number in all cases.

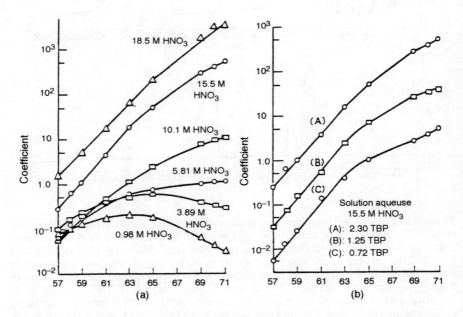

Figure 43. Change of partition coefficients with atomic number of the rare earths for (a) the extraction with pure tributyl phosphate at different concentrations of HNO_3, and (b) for different concentrations of tributyl phosphate diluted with carbon tetrachloride and with 15.5 M HNO_3 in the aqueous phase [3]

Table 8a. Aminepolycarboxylic acids

Name	Formula	Possible chelate rings
Nitrilotriacetic acid (H_3NTA)	$N(CH_2COOH)_3$	3
N-Hydroxyethylethylenediamietriacetic acid (H_3HEDTA)	$(HOCH_2CH_2)(HOOCCH_2)NCH_2CH_2N(CH_2COOH)_2$	4
Ethylenediaminetetraacetic acid (H_4EDTA)	$(HOOCCH_2)_2NCH_2CH_2N(CH_2COOH)_2$	5
1, 2-Diaminocyclohexanetetraacetic acid (H_4DCTA)	cyclohexane-$1,2$-diyl-N,N,N',N'-tetraacetic acid: ring-CHN$(CH_2COOH)_2$, ring-CHN$(CH_2COOH)_2$	5
Diethylenetriaminepentaacetic acid (H_5DTPA)	$(HOOCCH_2)_2NCH_2CH_2NCH_2CH_2N(CH_2COOH)_2$ with central N bearing CH_2COOH	7

Table 8b. Stabilities of some aminepolycarboxylic acid chelates of 2 °C

Cation	$\log_{10}K_{Ln(AA)}$				
	AA = NTA	AA = HEDTA	AA = EDTA[a]	AA = DCTA	AA = DTPA
Y^{3+}	11.48	14.65	18.09	19.41	22.05
La^{3+}	10.36	13.46	15.50	16.35	19.48
Ce^{3+}	10.83	14.11	15.98	–	20.5
Pr^{3+}	11.07	14.61	16.40	17.23	21.07
Nd^{3+}	11.26	14.86	16.61	17.69	21.60
Pm^{3+}	–	–	–	–	–
Sm^{3+}	11.53	15.28	17.14	18.63	22.34
Eu^{3+}	11.52	15.35	17.35	18.77	22.39
Gd^{3+}	11.54	15.22	17.37	18.80	22.46
Tb^{3+}	11.59	15.32	17.93	19.30	22.71
Dy^{3+}	11.74	15.30	18.30	19.69	22.82
Ho^{3+}	11.90	15.32	–	19.89	22.78
Er^{3+}	12.03	15.42	18.85	20.20	22.74
Tm^{3+}	12.22	15.59	19.32	20.46	22.72
Yb^{3+}	12.40	15.88	19.51	20.80	22.62
Lu^{3+}	12.49	15.88	19.83	20.91	22.44
Al^{3+}	–	–	16.13	17.63[a]	–
Fe^{3+}	15.87[a]	–	25.1	–	28.6
Co^{3+}	–	–	36	–	–

[a] At 20°C

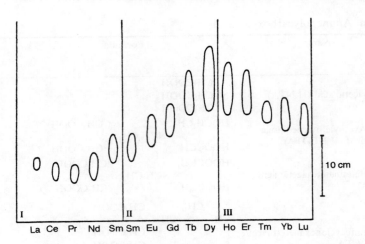

Figure 44. Rare earth chlorides developed for 1 week with ethanol–10% 2 N HCl placed in order of atomic number [2]. (Reproduced by permission of Elsevier Science Publishers BV)

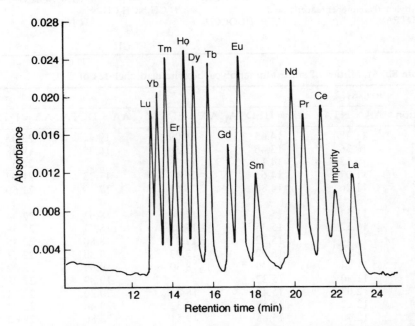

Figure 45. Separation of the lanthanides on Partisil-10 SCX. Experimental conditions: 25 cm × 4 mm i.d. column; sample: 10 µl of a solution containing 10 mg/ml of each lanthanide; linear programme from 0.005 M HIBA over a 30-min period at 1.3 ml/min and pH 4.6; detection at 600 nm after post-column reaction with Arsenazo I [4]. (Reproduced with permission, copyright (1979) American Chemical Society)

Although this would have appeared miraculous in 1940 we now have a wealth of separation methods. Ion exchange columns, ion exchange papers, thin layers, paper electrophoresis, continuous electrophoresis and liquid ion exchangers (in columns or impregnated papers) have all yielded analytically useful separations as long as suitable separation factors were worked out with complexants. Gas chromatography with volatile substituted acetylacetone as complexant also yielded excellent separations. Figures 45–48 show separations with some of the recent techniques: HPLC, capillary isotachophoresis, capillary zone electrophoresis and high speed counter-current chromatography.

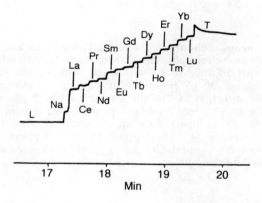

Figure 46. Isotachophoretic separation of lanthanides. L = 0.027 M KOH, 0.015 M α-HIBA, acetic acid, pH 4.92, 0.0025% polyvinyl alcohol; T = β-alanine [5]. (Reproduced by permission of Elsevier Science Publishers BV)

PREPARATIVE SEPARATIONS

Large scale separations were first achieved by Spedding and co-workers. One of his papers [8] can be considered a classic for preparative separations in general.

It became obvious in this work that elution chromatography permits only small loadings on a column and that overloading results in poor resolution. Displacement chromatography on the other hand permits the exploitation of the full capacity of a column. Typical separations of light rare earths by Spedding are shown in Figures 49 and 50. The cross-contamination at the zone fronts and rears is no problem, as these fractions can be recycled. Several 'rare earths factories' with column diameters of 50 cm and several metres long were working in the fifties and sixties, and supplied for the first time gram quantities of really pure rare earth elements.

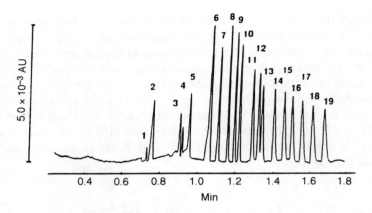

Figure 47. Simultaneous separation of alkali, alkaline earth and lanthanide metal cations by capillary zone electrophoresis. Carrier electrolyte, 10 mM Waters UVCat-1–4.0 mM HIBA (pH adjusted to 4.4 with acetic acid); capillary, 36.5 cm × 75 µm i.d. fused silica; voltage, 30 kV (positive); hydrostatic injection, 20 s from 10 cm height; indirect UV detection at 214 nm. Peaks: 1 = Rb (2 ppm); 2 = K (5 ppm); 3 = Ca (2 ppm); 4 = Na (1 ppm); 5 = Mg (1 ppm); 6 = Li (1 ppm); 7 = La (5 ppm); 8 = Ce (5 ppm); 9 = Pr (5 ppm); 10 = Nd (5 ppm); 11 = Sm (5 ppm); 12 = Eu (5 ppm); 13 = Gd (5 ppm); 14 = Tb (5 ppm); 15 = Dy (5 ppm); 16 = Ho (5 ppm); 17 = Er (5 ppm); 18 = Tm (5 ppm); 19 = Yb (5 ppm) [6]. (Reproduced by permission of Elsevier Science Publishers BV)

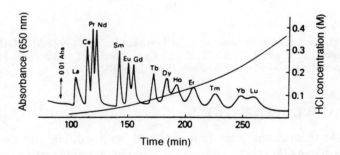

Figure 48. Gradient separation of 14 rare earth elements obtained by high-speed counter-current chromatography (HSCCC). The experimental conditions were as follows: apparatus, HSCCC centrifuge with 7.6 cm revolution radius; column, three multilayer coils connected in series, 300 m × 1.07 mm i.d., 270 ml capacity; stationary phase, 0.003 M DEHPA (di-(2-ethylhexyl)phosphoric acid) in n-heptane; mobile phase, exponential gradient of hydrochloric acid concentration from 0 to 0.3 M as indicated in the chromatogram; sample, 14 lanthanide chlorides each 0.001 M in 100 µl of water; revolution at 900 rpm; flow rate, 5 ml/min; pressure, 300 p.s.i. [7]. (Reproduced by permission of Elsevier Science Publishers BV)

The chromatographic factories have been superseded now by series of batch extractors working with 50 kg batches, quantities that were never

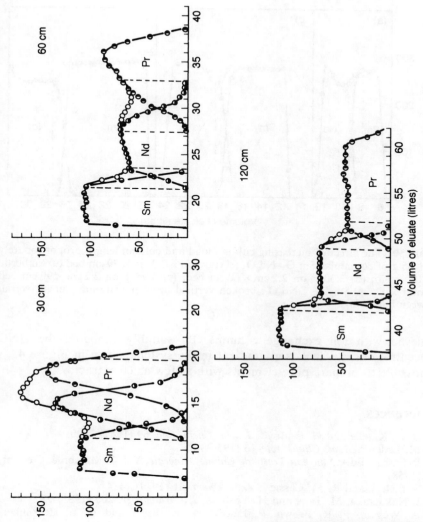

Figure 49. The elution of mixtures of Sm, Nd and Pr from −30+40 mesh size Amberlite IR-100 beds, 2.2 cm diameter and 30, 60 and 120 cm long, with 0.1% citrate solution at a pH of 5.30 and a linear flow rate of 0.5 cm/min. ○ total R$_2$O$_3$, ◐ Sm$_2$O$_3$, ● Nd$_2$O$_3$, ◐ Pr$_2$O$_{11}$; broken vertical lines indicate overlap between bands [8]

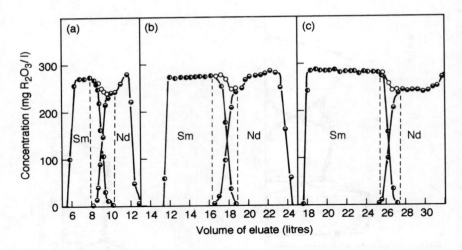

Figure 50. The effect of increasing column load and column length proportionately. (a) 1.65 g of equimolar Sm_2O_3–Nd_2O_3 mixture on a 2.2 cm × 30 cm bed of Amberlite IR-100 resin; (b) 3.30 g on 2.2 cm × 60 cm bed; (c) 4.95 g on 2.2 cm × 90 cm bed. ○ total R_2O_3, ● Sm_2O_3, ◐ Nd_2O_3; broken vertical lines indicate amount of overlap between bands [8]

achieved with ion exchange columns. By suitably changing the HNO_3 concentrations, i.e. in a step gradient in the system shown in Figure 43, it is possible to prepare pure elements with as few as 60 extractors.

References

[1] B. H. Ketelle and G. E. Boyd, *J. Am. Chem. Soc.* **69** (1947) 2800.
[2] M. Lederer, *Anal. Chim. Acta* **15** (1956) 46.
[3] P. Pascal (Ed.), *Nouveau Traité de Chimie Minèrale*, Vol. 7, Masson & Cie, Paris (1958).
[4] S. Elchuk and R. M. Cassidy, *Anal. Chem.* **51** (1979) 1434
[5] I. Nakatsuka, M. Taga and H. Yoshida, *J. Chromatogr.* **205** (1981) 95.
[6] A. Weston, P. R. Brown, P. Jandik, W. R. Jones and A. L. Heckenberg, *J. Chromatogr.* **593** (1992) 289.
[7] E. Kitazume, M. Bhatnagar and Y. Ito, *J. Chromatogr.* **538** (1991) 133.
[8] F. H. Spedding, *Discuss Faraday* Soc. **7** (1949) 214.

(ix) Chromatography at tracer levels

The English word 'tracer' is generally used to refer to radioactive isotopes either mixed with inactive isotopes or on their own. It is not a good word as its meaning is not clear. The French 'radioisotope sans entraineur' i.e. without an inactive carrier, is a bit better and the German term 'die Chemie der unwägbaren Mengen' i.e. the chemistry of amounts so small that they cannot be weighed, is perhaps the best. In radiochemistry tracer amounts are of the order of a thousandth of a microgram or less, depending on the half life of the tracer. Such amounts give several hundred to several thousand counts on a Geiger counter, enough for plotting an accurate chromatographic peak.

There are two elements, francium and astatine, of which the longest lived isotope has a half life of minutes and hours respectively. Then there are elements like polonium, which has some longer lived isotopes, but in weighable amounts they emit such intense alpha-radiation as to influence the stability of their valencies in solution. Their chemistry has been studied mainly by the use of two-phase equilibria, i.e. solvent extraction and ion exchange equilibria, paper chromatography and paper electrophoresis. In all these methods the detection and quantitation were performed by radioactivity counting. This of course limits the 'evidence' so obtained. Usually no crystalline compound or its analysis and no spectrum can support the chromatographic findings. These have thus still more the nature of conjectures than our chemistry with 'weighable' amounts.

Some aspects of the chemistry of polonium and of protactinium will be discussed here, as I have done some of it myself.

The chromatography of francium was investigated mainly by Adloff [1] and the chromatography of astatine has been reviewed in detail recently[2].

POLONIUM

Polonium was the first new element isolated from uranium and thorium minerals by Marie Curie in 1898.

Natural polonium is an alpha-emitter with a half life of 138.4 days In the radium decay series it is formed from $^{210}_{82}Pb(RaD)$.

$$^{210}_{82}Pb(RaD) \xrightarrow[22\ yr]{\beta} {}^{210}_{83}Bi(RaE) \xrightarrow[5\ days]{\beta} {}^{210}_{84}Po(RaF) \xrightarrow[138.4\ days]{\beta} {}^{206}_{82}Pb(RaG)$$

It can be isolated from solutions of RaD by spontaneous deposition on a silver foil and redissolution with HNO_3. A very good account of polonium chemistry can be found in the small book by Bagnall [3].

Although milligram amounts of polonium have been obtained, they emit such strong alpha-radiation that in solution they generate large amounts of H_2O_2 and thus autoxidation of lower valencies, even in 5×10^{-4} molar solution autoxidation occurs within about 10 minutes.

There is good evidence, however, that the 'usual' valency of Po in solution is 4, and in HCl solution Po exists in the form $H_2Po(IV)Cl_6$. The Po(IV) can be reduced to Po(II) in HCl by SO_2 or hydrazine in the cold. During autoxidation, the oxidation curve also suggests the presence of an intermediate valency of 3.

When Po (IV) is chromatographed on paper with butanol–HCl mixtures, it moves close to the liquid front like the strongly hydrophobic $HAuCl_4$, $HgCl_2$ and $TlCl_3$, and can be separated easily from RaD and RaE, i.e. from Pb and Bi, as shown in Figure 51.

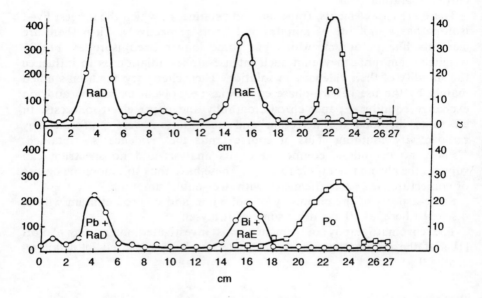

Figure 51. Top: paper chromatographic separation RaD-RaE-Po. Bottom: separation Pb–Bi–Po (Pb and Bi in macro quantities) [4]. Solvent: butanol saturated with 1 N HCl

Since tracer amounts of Po(IV) are reduced by SO_2 to the divalent state, we attempted chromatography in an atmosphere of SO_2. This gave fast-moving spots in butanol–HCl mixtures indistinguishable from those on chromatograms developed in air.

Polonium (IV) is also adsorbed on cellulose paper from aqueous HCl, but again the behaviour in a SO_2 atmosphere is identical to that in air. If Po(II) had been formed it could not be distinguished in butanol–HCl nor in adsorption from HCl solutions.

Thirty years ago a range of hydrophobic papers was commercially available. These were acetylated papers with different degrees of acetylation. Unfortunately these are not stable over long periods of time and hydrolyse to acetic acid and cellulose within several months.

As shown in Figures 52 and 53, there is a good distinction between chromatograms developed in air and SO_2 on acetylated cellulose papers. The reduced form moves near the liquid front, whereas the non-reduced form is adsorbed on the acetylated paper. As regards the possible inter-

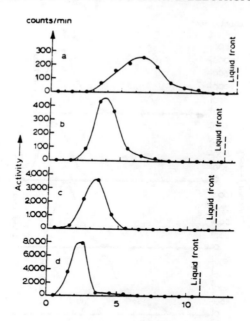

Figure 52. Activity distribution on chromatograms of Po(IV) chromatographed in air (or CO_2) on Schleicher & Schüll 20–25% acetylated paper. Solvents and R_F values: (a) butanol–6 N HCl (2:1), R_F 0.47; (b) butanol–8 N HCl (2:1), R_F 0.31; (c) butanol–10 N HCl (2:1), R_F 0.28; (d) butanol–12 N HCl (2:1), R_F 0.21 [5]; (Reproduced by permission of Elsevier Science Publishers BV)

mediate valency, i.e. Po(III), this seems to exist in chromatograms of solutions of Po(IV) to which cuprous chloride has been added before developing the chromatogram in an inert (CO_2) atmosphere: a peak which moves less than Po(II) and more than Po(IV) was observed. See Figure 54.

Since this work was done in 1960, better reversed phase TLC systems have been worked out which would also permit faster development, and thus less risk of oxidation during chromatography. The lower valencies of polonium should perhaps be re-examined with these improved means.

PROTACTINIUM

Like polonium, protactinium is a natural radioelement, and was first isolated from pitchblende in 1917. The long-lived isotope ^{231}Pa has a half life of 32, 480 years. It is not easy to work with, as its radiation characteristics are very inconvenient and it is about as toxic as plutonium, with a lethal dose around 8 micrograms. The UK Atomic Energy Authority

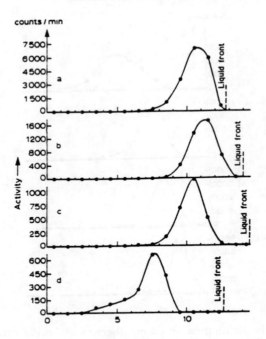

Figure 53. Activity distribution on chromatograms of Po (originally as Po(IV)) chromatographed in an SO_2 atmosphere on Schleicher & Schüll 20–25% acetylated paper. Solvents and R_F values: (a) butanol–6 N HCl (2:1), R_F 0.84; (b) butanol–8 N HCl (2:1), R_F 0.80; (c) butanol–10 N HCl (2:1), R_F 0.72; (d) butanol–12 N HCl (2:1), R_F 0.61 [5]. (Reproduced by permission of Elsevier Science Publishers BV)

extracted 125 grams of pure protactinium around 1960 and this seems to be the world's stock of this element.

For most chemical work the tracer ^{233}Pa, a beta-emitter with a half life of 27.4 days, is much easier to handle.

The stable valency of protactinium is five, just as with niobium and tantalum; but as it is bigger it is more metallic in nature. Protactinium (V) can be reduced to Pa(IV) in aqueous solutions, for example with zinc amalgam. In HCl below 6 N, Pa(V) has a tendency to hydrolyse. It forms very stable fluoride and mixed chloride–fluoride complexes in HF–HCl mixture which do not hydrolyse.

Paper chromatography with HF–HCl mixtures [6, 7]

In mixtures of organic solvents, e.g. butanol or acetone, with HCl and water, there is usually a tail on the paper due to hydrolysed species, as occurs also with Zr(IV) solutions.

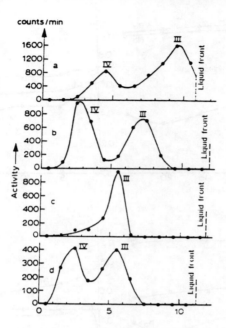

Figure 54. Activity distribution on chromatograms of Po(IV) mixed with Cu(I) chromatographed in a CO_2 atmosphere on Schleicher & Schüll 20–25% acetylated paper. Solvents and R_F values: (a) butanol–6 N HCl (2:1), R_F 0.41 and 0.88; (b) butanol–8 N HCl (2:1), R_F 0.23 and 0.60; (c) butanol–10 N HCl (2:1), R_F 0.46; (d) butanol–12 N HCl (2:1), R_F 0.21 and 0.49 [5]. (Reproduced by permission of Elsevier Science Publishers BV)

Good chromatograms are obtained in mixtures such as butanol–HCl–HF – water. Plastic development jars have to be used, of course. As shown in Figures 55 and 56, the R_F value of Pa(V) varies little with the HF or the HCl concentration and is usually around 0.4. Thus, Pa(V) is extracted into butanol but not as much as Fe(III), Nb(V) or Ta(V), and numerous separations are possible.

When we had just completed this work I met an eminent radiochemist, whose co-worker studied the liquid–liquid extraction of Pa(V) from HCl–HF media. He told me that he was about to formulate the explanation why Pa(V) is not extracted at all into organic solvents. Our results convinced him that he should try even more polar solvents.

The data shown in Figures 55 and 56 can be obtained in about two weeks and require altogether 12 sheets of Whatman paper. The speed and simplicity of flat bed methods has also induced several workers to use them in the field of quantitative structure–activity relationships (QSAR) in pharmacological research.

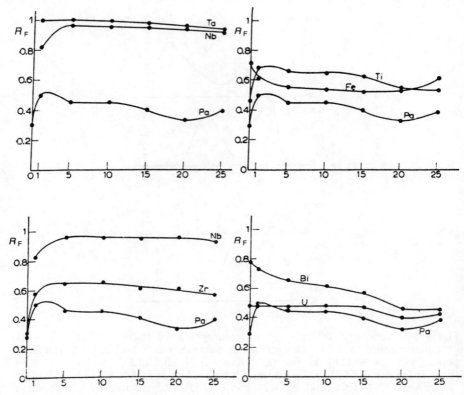

Figure 55. R_F values of metals in butanol–12 N HCl–conc. HF–water mixtures in the ratio 50:25:HF + H$_2$O 25. Above: the R_F values are plotted against the amount of HF in this mixture [7]. (Reproduced by permission of Elsevier Science Publishers BV)

The protactinate anion [8, 9]

Niobium (V) and tantalum (V) are sufficiently non-metallic to form niobate and tantalate in an alkaline solution.

Protactinium (V) is much more metallic, and the only hint towards a protactinate anion has been the observation that a hydroxide precipitated from aqueous solution loses several percent of its mass if washed with a reasonably concentrated KOH solution.

When a solution of ^{233}Pa(V) in HCl is evaporated in a beaker and 5 N NaOH is then added, the chromatogram of the resulting solution (or suspension) shows that all the Pa(V) has remained at the point of application. This can be interpreted in terms of an insoluble precipitate being present or a soluble species being absorbed very strongly on the paper (or its impurities).

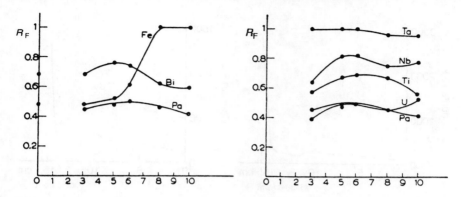

Figure 56. R_F values of metal ions in the mixture butanol–conc. HF–HCl (3–10 N) (50:1:49), plotted against the normality of HCl [7]. (Reproduced by permission of Elsevier Science Publishers BV)

When the Pa(V), after evaporation in the beaker, was fused with a few pellets of NaOH (or KOH) for a few minutes, and the cooled melt was diluted with water, the radioactivity was found to migrate with aqueous alkali as solvent but not as a compact single spot (Figure 57).

Here one should reflect whether this observation can be truly attributed to a 'protactinate' anion. We can only affirm that it behaves very differently from the sample treated with aqueous alkali (instead of fusion). Does movement on the chromatogram mean the existence of a PaO_4^- anion? Actually, it could also be due to the existence of one or more polyanions and/or complexes with silicate (from the walls of the beaker). Other workers repeated this work by fusion on a metal foil, and thus a reaction with silicate from the walls of the beaker is unlikely. Paper electrophoresis in 1 M KOH as electrolyte yields a single slow-moving anionic zone, much slower than phosphate or chromate. Does this suggest a polymeric anionic species or very strong ion pairing?

Equilibrium experiments with Dowex 2 anion exchange resin gave the same 'slope' as phosphate tracer (^{32}P) (see Figure 58), suggesting that the anion formed with Pa(V) could be triply charged (assuming that only a single species exists).

A number of electrophoretic experiments aimed at reducing possible adsorption or precipitation reactions all yielded the same slow anionic band.

So what can be deduced from such 'tracer' studies? Firstly, that aqueous alkali has little and alkali fusion has a strong effect in producing a form soluble in aqueous alkali. If we had only the anion exchange equilibrium data we would conclude that a triply charged species is formed. However, this would assume that there is a single species, and we cannot rule out the existence (or an equilibrium) of several species. The electrophoretic movement is too slow for an anion like PO_4^{3-} or CrO_4^{2-}.

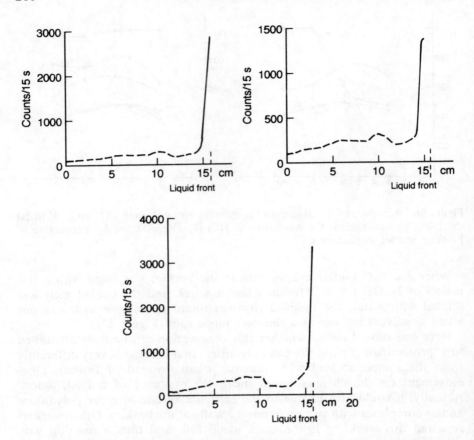

Figure 57. Paper chromatograms of 'protactinate' developed on washed Whatman No. 1 paper with 1 N, 2 N and 3 N KOH. Note that most activity is on the liquid front but some remains behind [9]. (Reproduced by permission of Elsevier Science Publishers BV)

So, except for the evidence that Pa(V) is solubilized in an alkaline fusion, the rest is at best conjecture.

A biochemist once stated correctly that 'chromatography is the fastest method for drawing wrong conclusions'. This may be so, but if one does enough experiments one finds it difficult to draw any conclusions at all.

References

[1] *Gmelin Handbook of Inorganic Chemistry*, 8th edn., 'Francium', Springer-Verlag (1983).
[2] *Gmelin Handbook of Inorganic Chemistry* 8th edn., 'At Astatine', Springer-Verlag (1985).

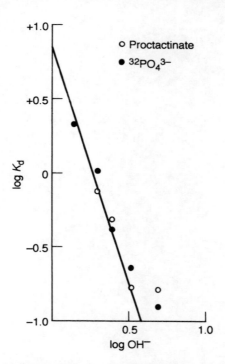

Figure 58. Anion exchange adsorption of protactinate and tracer $^{32}PO_4^{3-}$ in KOH of varying concentrations; $-\log K_d$ is plotted against $\log OH^-$ of the solution [9]. (Reproduced by permission of Elsevier Science Publishers BV)

[3] K. W. Bagnall, *Chemistry of the Rare Radioelements*, Butterworths, London (1957).
[4] M. Lederer, Doctoral Thesis, Paris (1954).
[5] B. Klinken and M. Lederer, *J. Chromatogr.* **6** (1961) 524.
[6] M. Lederer and J. Vernois, *Compt. Rend.* **244** (1957) 2388.
[7] J. Vernois, *J. Chromatogr.* **2** (1959) 155.
[8] Z. Jakovac and M. Lederer, *J. Chromatogr.* **1** (1958) 289.
[9] Z. Jakovac and M. Lederer, *J. Chromatogr.* **2** (1959) 411.

INDEX

adsorption, on alumina 8
adsorption paper chromatography 43, 44
alkalies, paper chromatography 38
alumina 102
　adsorption on 8
ammonium phosphomolybdate 108
anions, paper chromatography 41, 42, 43
antimonic acid 109
apparatus, for paper chromatography 29
arsenate 166
arsenite 166
astatine compounds, gas chromatography 138

boric acid 162
bromate 176
bromochloro-osmates(IV) by HPLC 124
bromochlororhenates by HPLC 125
bromohydroborates 163

capillary analysis 8
capillary zone electrophoresis 74–77
carrier-free tracers, paper chromatography 39
carotenoids 4
'charge', on an ion 111–114
chlorobromorhenates(IV) 181
chlorohydroborates 162
chlorophylls 3
chromium(III) thiocyanate complexes 114, 115

cobalt(III)ammine complexes 115
cobalt(III) complexes,
　reactions studied by HPLC 119–124
　TLC of 46, 47
condensed phosphates 163, 166
condensed phosphates–arsenates 167
continuous electrophoresis 54, 59
coordination compounds, synthesis 39
Craig machine 15, 16, 17, 18
cross electrophoresis 65
cyclic polyphosphates 169

definition of chromatography 1
displacement 3, 7
displacement chromatography, or rare earths 205

electrophoresis 50–79
　in fused slats 66, 67, 68, 69
elemental sulphur 172
elution 3

fast chromatography of radionuclides 119, 120
fluorochloro-osmates(IV) by HPLC 124, 125
frontal analysis 4, 7

gas chromatography 133–139
gel electrophoresis 65, 66, 67
gel filtration 80–85
geochemical prospecting 38

halogenoborates 162
halogen oxyacids 176

heteropolyacids 105
hexachlororhenate(IV) 182
high performance paper
  electrophoresis 77, 78
high voltage paper electrophoresis
  56–58
history 1
HPLC 118–125
  historical 118
  separation of anions 119
hydrogen isotopes, separation 140,
  141, 142

inorganic gases 135
inorganic ion exchangers 101
inorgano-organic ion exchangers 106
iodate 176, 178
iodohydroborates 163
ion chromatography 126–131
  separation of alkalies 128
  separation of anions 127, 128
ion exchange 86–117
  celluloses 97, 98, 99, 100
  data 93, 94
  in organic solvents 96, 97
  papers 106
  resins 86, 87
  techniques 89, 90, 91, 92
  thin layers 106
ion exchangers, titration curves 87
iridium(III) chloro-aquo-complexes 48
iridium(IV) chloro-aquo-complexes 47
isotachophoresis 70–75
isotope separations 140–150
  by ion exchange, 144–148

liquid ion exchangers 98, 99, 101
lithium isotopes, separation 143, 144

mercury, by gas chromatography
  136
metal halides, volatile 135, 137
metal hydrides, volatile 135, 136
metal oxides, volatile 136
mixed ligand complexes, paper
  electrophoresis 61
mobilities of metal ions in paper
  electrophoresis 53–57
molybdate and its thia-substituted
  derivatives 130, 132

nitrogen isotopes, separation 142,
  145, 149

optical isomers 151–161
  of Co(III) complexes 150–161
  of Cr(III) complexes 158, 159
outer sphere complexes, paper
  electrophoresis 61
paper chromatography 28
  apparatus 29
  historical 50
  techniques 51, 52
partial constants, R_M values 23
partition chromatography, history of
  10, 19
perbromate 176
perchlorate effect 96
periodate 176
phosphorus oxoacids 164
platinum(II) complexes 40, 44
polonium 209
  lower valencies 210, 211, 212
polyphosphates 163, 165
  by ion chromatography 129, 130
polysilicate ions 83
polysulphanes 175
polysulphides 175
polysulphondiphosphates 172, 173
polythionates 60, 170, 171
  paper chromatography 43
potassium isotopes, separation 145
preparative paper chromatography 40
protactinate 215–217
protactinium 211–215

R_F value(s) 20
  calculated, of peptides 26
  conversion to R_M values 22
  of inorganic ions 30, 31, 32, 33, 34,
    35, 36, 37, 38
R_M value 20
rare earths 199–205
rare earth–EDTA complexes 62, 63
reversed phase partition
  chromatography 19
rhenium 180
  valencies, separation 180, 181
rhenium(IV) bromo-chloro complexes
  45
rhodium 193
  complexes with pyridine 198
  complexes with stannous salts 197
  polymeric ions 81
  solution chemistry, in hydrochloric
    acid 194
  in oxalate solution 197

INDEX

rhodium perchlorate 194, 195
rhodium(III) sulphate complexes 197
ruthenium 187
   polymeric ions 81
   solution chemistry, in hydrochloric acid 187
   in nitric acid 187–193

salting out, in paper chromatography 37
   in solvent extraction 14
selenium, by gas chromatography 137
selenopolythionates 171
separation of proteins by two-dimensional electrophoresis 25
separations on ion exchange columns 95, 96, 98
Sephadex gels 81
silica 101
silicate, mono- ion 83
silicic acid 83–84
sodium isotopes, separation 143
solvent extraction 10
   with ether,
      metal bromides 11
      metal chlorides 10, 11
      metal fluorides 12
      metal iodides 12
      metal nitrates 13
      metal thiocyanates 13
structures, schematic of ion exchange resins 88, 89
sulphur homocycles 174

technetium 181
   valencies 184, 185, 186
technetium(V) 182
theoretical plate concept 24
theoretical plates, of different methods 25
thin layer chromatography 28, 44
titanium dioxide 103
tracer level chromatography 208
tungstate, and its selena-substituted derivatives 131
   and its thia-substituted derivatives 131
Tswett 2

uranium isotopes, separation 145–150

volatile metal complexes 136

zirconium, polymeric ions 80, 83
zirconium oxide 102
zirconium phosphate 104
zirconium tungstate 108